preserving our roots

For Poppy,

Who is an absolute inspiration
to me in saving our heritage
foods, flavors and stories

Christina Melker
2019

THE SOUTHERN TABLE

Cynthia LeJeune Nobles, Series Editor

preserving our roots

MY JOURNEY TO SAVE SEEDS AND STORIES

JOHN COYKENDALL *with* CHRISTINA MELTON

PHOTOGRAPHS BY SARAH HACKENBERG

LOUISIANA STATE UNIVERSITY PRESS   BATON ROUGE

Publication of this book is made possible in part by the support of the
John and Virginia Noland Fund of the Baton Rouge Area Foundation.

Published by Louisiana State University Press
Copyright © 2019 by Louisiana State University Press
All rights reserved
Manufactured in Canada
First printing

Designer: Barbara Neely Bourgoyne
Typefaces: Ingeborg, text; Verb, display
Printer and binder: Friesens Corporation

Library of Congress Cataloging-in-Publication Data
Names: Coykendall, John, 1943– author. | Melton, Christina, 1970– author.
Title: Preserving our roots : my journey to save seeds and stories / John
    Coykendall with Christina Melton ; photographs by Sarah Hackenberg.
Other titles: Southern table.
Description: Baton Rouge : Louisiana State University Press, [2019] | Series:
    The Southern table | Includes index.
Identifiers: LCCN 2019017924 | ISBN 978-0-8071-7036-6 (cloth : alk. paper)
Subjects: LCSH: Seeds—Louisiana—Washington Parish. | Seeds—
    Harvesting—Louisiana—Washington Parish. | Seeds—Storage—Louisiana—
    Washington Parish. | Heirloom varieties (Plants)—Louisiana—Washington
    Parish. | Cooking.
Classification: LCC SB118.38 .C69 2019 | DDC 631.5/210976311—dc23

My passions in life have always centered on farming, food, storytelling, and art. These interests are interwoven like the vines of the peas I nurture and grow. Their convergence has led me to pen hundreds of journals over my lifetime, books that are filled with invaluable advice I've received about living, love, generosity, and gratitude. My greatest devotion is, of course, to my wife, Ashley, whom I married in 2017. This book is dedicated to her and to the people of Washington Parish, Louisiana, particularly the Vise family. I like to say the Vise family opened every door to me in this part of the country and "made all things Washington Parish possible." They have enriched my life in ways I cannot ever fully express.

—JOHN COYKENDALL

CONTENTS

Foreword, *by Mary Celeste Beall* ix

Acknowledgments xi

Introduction 1

Seasons of Life 11

Winter 53

Spring 91

Summer 123

Fall 155

Starting Again 189

Recipe Index 193

FOREWORD

It is such a blessing to work and live among artisans who value their craft and keep tradition and history alive within it, artisans like John Coykendall. So often in today's world, we lose the joy of the process. We think about the end result and forget to cherish the details and practices that get us there—not John.

John's efforts to document and share stories and traditions have created an invaluable collection of knowledge to pass down. He knows and teaches the satisfaction that comes from pulling dirt by hand to perfectly place a seed, tending to and caring for a plant as it grows, and even reviving plant species that would otherwise be forgotten.

Having grown up with a love of farming and gardening, John joined the Blackberry Farm team in 1999, bringing with him a wealth of knowledge in farming and farm culture. Blackberry Farm is dedicated to sharing the customs of southern hospitality and creating unique opportunities for our guests to learn, enjoy, and connect with the incredible land and history of East Tennessee in the foothills of the Great Smoky Mountains. In partnership with the Seed Savers Exchange, John uses land at Blackberry Farm to nurture and save dwindling heirloom seed varieties. He shares the techniques, knowledge, and stories that he has collected with the Garden Team and our guests in hopes that younger generations will carry on the traditions he cherishes.

From hand-sharpening his pencil with his pocketknife blade for journaling to pulling up a rocking chair by the Garden Shed's woodburning stove for storytelling, John embraces his role of keeping history alive. His attention to detail and passion for what he loves make his stories compelling and enjoyable for all who have the chance to interact with him.

I am so pleased that this book opens up an opportunity for even more people to learn and treasure the information that John imparts. It is truly a privilege to support his work and watch him share his passion and efforts with guests, and now readers, from all over the world.

Each year for the last two decades, we have watched John travel down to Louisiana to continue his journey through the world of heirloom seeds and southern farming culture. Each year he returns to Tennessee and Blackberry Farm with more stories, more friends, and of course, more seeds!

I am thankful that my children have grown up in an environment where they could run over to the garden, interact with John, pick vegetables straight from the ground, and embrace the enchantments of simpler times. John's stories and lessons have equipped them with valuable knowledge of tradition and growing—even when they were unaware they were learning—that they will one day pass on to their own children. Making fresh tomato sandwiches, shelling baskets of peas around the fire, digging up potatoes, and plucking the first summer strawberry from the vine are the kinds of moments my children will carry with them as fond memories of enjoying life's humble pleasures.

At Blackberry Farm, we treasure the opportunity to share the history and practices of this region and its people with all of our visitors. John is a conduit for capturing the spirit and magic of everyday life and turning it into a beautiful story. The material he has dedicated so much time to preserving maintains timeless value.

Carrying on a legacy is a great responsibility. We share John's feeling of obligation to ensure that the seeds, culture, heritage, and stories are entrusted to future generations.

Whether you're a gardening enthusiast or simply enjoy a great story, I assure you, you will find yourself learning and laughing through these pages. I hope this book reminds you to slow down and cherish small moments. These pages are filled with stories, people, and seeds that are easily forgotten, but all too important to forget.

—MARY CELESTE BEALL
Blackberry Farm Proprietor
Walland, Tennessee, January 2019

ACKNOWLEDGMENTS

I owe tremendous thanks to my co-writer and friend, Christina Melton. Without her, the stories in this book would likely have remained tucked away in the disparate pages of my journals, stored in the nooks and crannies of my Tennessee home. Thank you for making sense of them all and helping me to share them. I am sincerely grateful to Sarah Hackenberg for her creativity and beautiful, vibrant photographs that bring these stories to life, and to the expert editorial team at LSU Press, particularly Cynthia Nobles, James Long, and Catherine Kadair. Stan Ivester did an excellent job with the copyediting. Thank you also to my many families: at Blackberry Farm, especially Michael Washburn and the Beall family; in Washington Parish, particularly members of the Vise, Lang, O'Bryant, and Graves families, some who did not live to hold this book in their hands but whose stories live on in its pages; and finally, to my wife, Ashley, for her love and patience.

preserving our roots

INTRODUCTION

When I was a young man growing up in Tennessee, I wanted to learn everything I could from people who were older than me, people I viewed as the experts on living. I was studious in making note of every facet of life that they imparted to me. I saw an urgency in documenting every detail or sage bit of advice I could, before these old-timers died away. And I managed to record literally volumes of information about how earlier generations survived off the land and their own toil. But over the years, a strange thing happened. My late friend Homer Graves, a Louisiana farmer, explained it this way:

"I asked my best friend the other day, 'Where are all the old people?'"

My friend replied with a laugh, "We are the old people."

And I said, "How about that, I don't feel old. When did that happen?"

Increasingly, I find myself asking that same question. Now that I am one of the "old people," I am faced with a more pressing question. What will I leave behind? Our days here are numbered. I think it is a natural human desire to want to leave a legacy. I have tried to document the legacies of people I've met and befriended, so that they will live on. Their stories, now in my hands, come mostly from an off-the-beaten-path farming community in Louisiana, in rural Washington Parish. To me, Washington Parish represents a time capsule of what life was like in most of America up until the beginning of the last century. It is a microcosm of an agrarian culture that is in danger of being lost. The stories told to me by these old-timers are not extravagant, but rather of simple lives, richly lived. Many of these memories are in this book, which I have written in hopes that I will plant the seeds of a meaningful contribution of my own.

Just as I am motivated to save stories, I feel called to find and save rare seeds from food crops we once grew. We are in a race against time to preserve these vanishing varieties, and to preserve the knowledge of how to grow them and their surrounding culture. In many estimates, more than 94 percent of the fruits and vegetables we grew just one hundred years ago are extinct or are what is termed "functionally extinct." This means that while a certain variety of bean may exist somewhere, like say, your great-aunt's freezer, because no one else knows about it or has any way to reproduce it, it is steps away from extinction.

I once heard it described in this way. If you had one

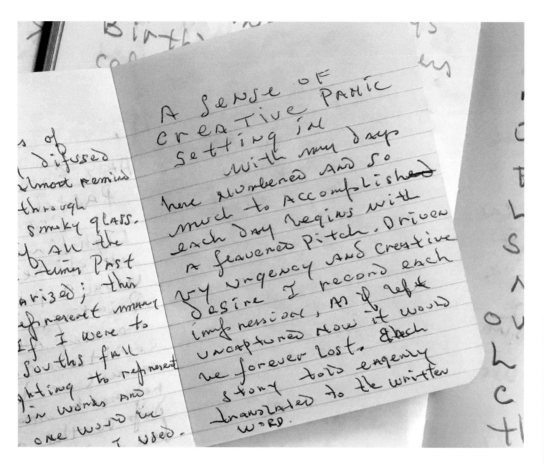

An October 27, 1993, journal entry capturing the "sense of creative panic" and urgency I feel in my work recording stories and collecting seeds in Washington Parish, Louisiana.

thousand rooms and one thousand keys, one for each room, which keys could you afford to throw away? I think that is a great comparison. Of all the seeds we have left, which ones can we afford to lose? The more seed varieties we lose, the more dependent we are on the few that remain, and the more vulnerable those become to extinction. Without a secure foundation of diverse varieties of seeds, our food sources are threatened by everything from disease, climate change, natural disaster, and failed stewardship, to global and political instability, and even economic greed.

Stacks of my journals from four decades of
visits to Washington Parish, Louisiana.

I have saved more than five hundred varieties of these seeds in my lifetime, mostly of beans, peas, and corn. I've done my best to document as much history about each seed as I can—where it originated, who grew it, how it was cultivated, and the best means of preparing it for the table. Not only do I enjoy saving these seeds, but I enjoy cooking and eating them too. This book features many of the recipes and cooking techniques I have learned from generations of southern farm cooks about how they served these old varieties to their families—and, of course, I was no stranger at their family tables!

Decades ago, when I started recording these stories and collecting this material, I wasn't thinking in terms of writing a book or even doing anything special with it. I was simply writing down bits of information that I found interesting. Years and years went by, and my collection of seeds and stacks of journals grew and grew; I wasn't sure what to do with everything I had saved.

It didn't hit me until I was visiting with my Washington Parish friend, Terry Seal, and she was looking through some of my journals. She looked me in the eye and said, "Now John, you do something with these books and all this work. Don't let this get away from you. Don't just let it gather dust on your shelf and nothing ever come of it. Put this out there where it can be preserved and people can read it and see it and learn from it." That's when I felt a little tinge of guilt and realized I was probably in real danger of letting my life's work slip away from me. I am good at recording this stuff, but I am not good at pursuing methods of getting things out. That's where I needed help.

I had no idea what to do with all this material or how to organize it. Then in 2015, I had the good fortune of meeting a documentary filmmaker from Louisiana named Christina Melton, who visited Blackberry Farm,

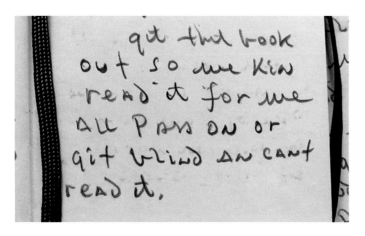

Friends often encouraged me to publish my journal notes in a book, including this admonition from one of my older acquaintances.

the renowned Tennessee resort where I work. We struck up a conversation about the Bayou State, and I showed her a few of my journals. She started the wheels rolling. In a joint effort, we began working on this book and a documentary film. Now seeing the potential of this project come to fruition, I have found this is the most dedicated I've been to any endeavor in my entire life.

In 2016, over the course of making the documentary, we realized the story could have an impact and appeal beyond this small Louisiana community. The story of my farming experiences in Washington Parish, Louisiana, as well as documenting and preserving its seeds, old farming ways, and the heritage and foodways of a bygone era, is, in a sense, about every farming community in America. All across our great nation, there is a growing desire to reclaim those traditional sensibilities and their connections to our food.

It is one thing to save endangered seeds and stories, but unless they are shared, they, too, are in danger of being lost forever. This book contains only a small fraction of the information I have collected over the years. It shares many of my favorite oral histories, folktales, farming techniques, recipes, and sketches taken from my journals. You'll learn about my work preserving heirloom varieties of food crops that were traditionally grown by farmers in the Deep South, and also in the Appalachian region of Tennessee, where I've lived and worked most of my life.

I also tell the stories of a past generation of farm life, told to me by people who lived and experienced it. My primary source of this material was a small circle of men and women who became close friends over the decades. My journal notes have been accentuated by chance encounters I've had with countless others I met over the years. The cadence and dialect of these voices play constantly in my ears like a favorite melody, and I have done my very best to capture the poetic manner in which the people I've interviewed express themselves. These tales focus on the fields and kitchens of this insular community in Washington Parish. This work is not intended to be an academic study or official record of this important farming district. Rather, it is a cultural legacy of agrarian life, once common across America, gleaned from front-porch visits, ranging from the subjects of planting and growing, to old-time medicinal remedies, cultural customs, and childhood memories. Like the growing seasons that dominated life on Washington Parish farms for over a century, this book reflects life in times past in every American rural community and farm, shaped as it was by the seasons of the year.

This project began when I met filmmaker Christina Melton, who did a documentary on my work for Louisiana Public Broadcasting in 2015. She spent time filming at my home in Tennessee, including a segment about my seed freezers, where I keep hundreds of seeds I have collected. She has also been invaluable in helping me write this book.

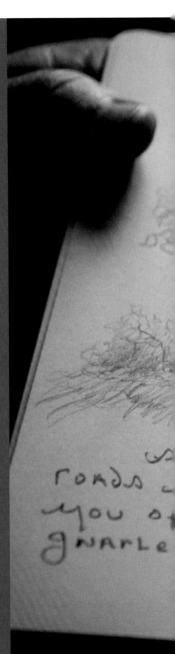

from life & experiences
has developed and
especially an Appreciation
of how Precious life is.
"You Don't realize
how VALUABLE no Precious
life is until you're
Down towards the end of
it. Our Destiny is in
his hands and only he
Knows what that is. We
may think we know,
but God alone knows
what our Destiny is.

30 oct. 98

OUR Destiny

many times as we
were sitting by the
car shed and having
Discourses on the ways
of life, the subject
of our destiny would
Come up. After
having lived Nearly
87 Years, a lot of
insight And wisdom

An October 30, 1998, journal entry describes one of many of my visits with my Washington Parish friend, Seldon Lang. [The entry begins on the righthand page and continues on the facing page.]

Facing page: Sketches made on Washington Parish back roads.

"I pict ah many
ah vol ah cotton
vack in them days
hun, ah many ah vol,
I shor did"
MR. arlie O'Bryant

ong Dusty gravel
d red clay vanks
en all Short
vack jack OAKs

SEASONS OF LIFE

The thing I love best about seeds is watching them grow. I remember my good friend Homer Graves, from Washington Parish, Louisiana, always talking about his fascination with the same thing. He would tell me, "I love to go out after a rain, about two days after you've planted some peas. You see that little green crook come up above the surface of the soil, just before the seed leaves pop out."

I know exactly what he is talking about. I love to watch seeds sprout. In a way, they are like people. New leaves stretch open and turn to face the sun as it moves across the sky. Vines spread and grow thick to adapt to their surroundings, winding and climbing up any structure they find. The old farmers I talk to called peas and beans "a 365-day-a-year crop." As Mr. Homer would say, "You could get three makins out of 'em." In other words, you can plant them three times over a season and, for farm families of past generations, those seeds were a mainstay that could stave off hunger.

For well over a hundred years, right after the last spring frost, farmers across the South would plant field peas, beans, and butter beans. Not long after the first tender sprouts emerged, the plants grew rapidly, and they would start producing. First came green field pea "snaps," the slender, delicate young bean pods so named because of the sound they make when you break them. Then came immature green peas and beans from tender pods that you shelled and ate fresh.

Soon, the plants would bear pods so fast you couldn't keep them picked through the summer and into the fall. As the season turned cooler, pea and bean pods were often allowed to dry on the vine, where they could be easily collected and shelled on front porches or by the fire after supper. All winter long, dried beans and peas would be used in hearty soups or soaked and smothered with smoked meat and seasoning. Once spring arrived, you started the cycle all over again.

Nothing went to waste. You always saved enough to eat and to serve as next year's seed stock. Farmers turned livestock out into fields of dried pea vines for winter grazing, and pea hulls made good kindling for fires. With their nitrogen-dense roots and organic matter, these legumes continuously rejuvenated the soil throughout the growing season. I remember some old-timers recalling their parents' assurances, "Son, if you got some butterbeans and peas and some meal to make some cornbread or cornpone, you're not going to starve. You will make it."

Right: As a child, I loved working with my father in our family garden. *Photo from author's collection.*

An October 14, 2007, journal illustration of one of my good farmer friends, Seldon Lang, working in his fields.

Of course, these seeds carry with them much more than their potential as food to sustain human beings. They represent the living history of the people who cared for and tended to them, cultivated them, and passed them down. The stories of these people intertwine, like pea vines, with details of how our ancestors grew the seeds, prepared them, safeguarded them, shared them between growing seasons, and served them at the table. These seeds represent an unbroken chain of our collective heritage and traditions, with no missing links, a gift our forbearers have handed down to us.

As the recipient of that gift, I believe I have a sacred responsibility to pass those seeds and the stories they carry on to future generations, and I hope to encourage others to do the same. If we do not, we are the break in the chain. Extinction is forever. Once those seeds and stories are gone, they are gone. I have felt this sense of obligation very deeply since I was a teenager. Now that I am up in my seventies, I have a growing sense of urgency, especially since the cultural heritage of cultivating your own food is in jeopardy of being lost.

I was born in 1943 in the foothills of the Smoky Mountains of Knoxville, Tennessee, where I was also raised until adulthood. My grandfather was a US congressman, my father a banker, and my mother a schoolteacher. All of these family members loved gardening,

and they planted that seed in me at an early age. When I was about ten years old, my dad showed me how to plant corn and potatoes in a little plot beside our garage. I was thrilled when I planted and then dug my very first new potatoes out of the ground. The skins were so tender, they would just about come off when you washed them. They were so delicious! Not like the ones you buy at the store today.

I have a similar memory from my grandfather's farm outside of Knoxville, growing Purple Top Turnips. I was amazed at how beautiful the roots looked, with their creamy white bottoms and purple-blushed crowns. I like to think back to one particular winter day, when I pulled some and took them home to my mother. She cooked turnip roots in the simplest way, thick-sliced and steamed, and seasoned with only butter, salt, and pepper. They had a heavenly taste I vividly remember and love still.

Food can do that—especially homegrown food. It can magically transport us to a different place and time and surround us with the comfort of loved ones who are far away, or with those who may no longer be with us. Even the simplest foods link us to a history, a culture, and a way of life uniquely ours and collectively shared. Smells, tastes, and textures of certain foods prepared certain ways remind us of our ancestry, our youth, our

My first recollection of a fall garden and turnips dates back to a fall day sometime in the early 1950s, when my grandfather took me to the farm and we pulled purple-top turnips from a fenced-in patch where a hog had turned in to fatten. I remember how big and ferocious that hog looked, and how fascinated I was by the beautiful purple tops of the turnips. After returning home that evening I insisted on having the turnip roots for dinner. They were served with butter, salt, and pepper and have remained a favorite ever since.

family kitchens, friends, neighbors, and cherished times together. Those memories are an important part of the story of who we are, and they enhance our lives.

Research shows that flavor impressions register in our brains from the earliest stages of childhood, and even before we are born. Most of our sense of taste is largely impacted by our sense of smell. That is why certain flavors and aromas can evoke such strong memories, whether it is summer tomatoes grown in your grandmother's garden, the taste of tender shelled butter beans, or hot buttered cornbread fresh from your mother's oven. Nothing seems to match those tastes as adults. This is not our imagination. For the last half-century, flavor has been bred out of much of the food we buy by the use of processes that make it prettier and easier to ship, store, and serve. Reclaiming those flavors that we remember is essential to preserving our collective cultural heritage, a heritage that will be lost if we don't scramble to save it. My mission is to pass that legacy on to generations who don't know what they are missing.

Some of my most cherished memories have come from meals shared around tables and in kitchens, learning everything I could from old cooks I've met. Most of them talk about eating "in season," and preserving what you couldn't eat when produce was plentiful, to save it for a time when there might not be food. My good friend Beulah Mae Lang, also from Washington Parish, told me earlier generations ate what they grew. They didn't have a lot of variety, but there was no shortage of creativity when it came to preparing and seasoning delicious meals. It was farm to table by necessity. It was a pretty simple diet, but it was a hearty, flavorful, and healthy one.

I cannot recall a time in my life when I wasn't irresistibly drawn to growing things. I have farmed for most of my adult life. For the past eighteen years, I have had the privilege of working as master gardener at the famed Blackberry Farm, in Walland, Tennessee. Blackberry Farm is part of the Relais and Châteaux international collection of individually owned and operated resorts. This enchanting property has been celebrated in magazines and films for its luxury and service, for its beautiful natural setting, for its lush organic gardens, and for its world-class, farm-fresh cuisine.

At Blackberry, I have had a heavenly place to cultivate many of the old seed varieties I have collected, and to have them appreciated and prepared by some of the world's finest chefs. One remarkable thing is that older vegetable varieties have such superior flavor, and that incomparable taste is what we want to pass on to future chefs, cooks, and young families. I have also had a soapbox of sorts to quietly proselytize about what I call "the seed gospel." It is my belief that seeds and their stories simultaneously link us to our ancestors and our descendants, ground us to the land on which we live, and nourish our bodies and spirits.

In the last half-century, we've become disconnected from the sources of the food we eat. There was an old fellow down in Louisiana I was talking to one time, and he asked what I did up at Blackberry Farm. I told him that I grow food for the resort's restaurants. "It's farm to table," I explained. He looked at me funny and said, "Farm to table? When we were growing up it was, 'farm to table' or starve!" I know what he meant. In generations past, it literally was farm to table every day. People were intimately connected to their farm roots and to the food they ate. Today, the idea of "farm to table" eating is a modern trend that actually looks backward to our roots.

One time two visitors came to Blackberry Farm, a lady and her son. The young man had hated tomatoes all his life. I never could figure how someone would not love a good homegrown tomato. That day it finally dawned

The 1913 *William Henry Maule Seed Catalogue* that I discovered as a teenager was a treasure to me, inspiring my life's work in searching for heritage seeds and in preserving them through art.

on me. That boy had never eaten a real tomato. Like many people, he'd only had mass-produced tomatoes. To me, these commercial mega-farm tomatoes have plastic skins and Styrofoam interiors, and they smell like water. So, this boy came down to a tomato tasting we had set up to showcase the heirloom varieties we were growing, and I finally talked him into tasting one. Next thing I knew, he sat down and cleaned off that whole cutting board himself. He couldn't get enough tomatoes.

Another visitor to Blackberry Farm was a lady celebrating her 106th birthday, but who looked and acted like she was in her 70s. At the time, I was growing some old Ponderosa tomatoes, an heirloom variety that was a forerunner of modern beefsteak tomatoes. She was as sharp as can be, and I asked her if she remembered the old Ponderosas. She said, "Oh my dear, I haven't thought about that tomato in 85 years. They were delicious." I finally found someone who was actually around to experience an old variety when they were still being extensively grown. It is that kind of revelation that inspires me to try to bring back these older seeds and the flavor experiences and cultural heritage they represent. As caretaker of these varieties, I never take for granted that any particular heirloom seed or bit of farming knowledge from the past will be there next year.

My first real interest in seeds and seed saving began when I was sixteen. I was exploring an abandoned railroad station west of Knoxville, Tennessee, where I found a *William Henry Maule Seed Catalogue* from 1913. This trade catalog was in pristine condition and was filled with intriguing varieties of vegetables and fruits I had never heard of. Finding that catalog was transformational, both in terms of my interest in seeds and in art. The illustrations were a combination of traditional botanical prints and fine engravings, and they provided exquisite details of old tomatoes, beans, peas, and other vegetables. The images were fantastic, but I immediately wondered what had happened to those seeds. Were they still around? Could you still grow them? Since that day, it has been my lifelong quest to track them down.

Over the years, I have collected as many old varieties from the *Maule Catalogue* as I can find, as well as other legacy varieties I've heard about through my searches. I store my seeds in three freezers in my basement. Rummaging through my collection is almost like going on an archeological dig. One fact that often strikes me when I look in my freezers is how many of the people who gave me those seeds are now deceased, and that each may well have been the last to have any given variety. I don't know that for sure, but I treat every seed I receive as though I was the last person who has it. As I advance in age, I hope I live long enough to grow out all the seeds I've acquired.

Many of the seeds I have are preserved, registered, and offered through the Iowa-based Seed Savers Exchange, the world's largest nonprofit organization maintaining genetic plant diversity. Safeguarding and sharing seeds with people who will grow them is the best way to ensure their future. Saving the history of the seed is also an important piece of the puzzle. Say, I have a handful of red beans, and I show them to someone who might say, "Oh, those are nice. What else do you have?" But their perspective changes if I tell them some historical details about the seed, like how it got to America and biographical details about who grew it, and where and in what special conditions. Perhaps I could share popular methods of cooking the variety or someone's childhood memory of enjoying them at the dinner table. Those stories anchor you in a place and time. Seeds by themselves are nice and may taste good, but it is the story that gives you a connection to how they fit into the history of our lives.

To record seed history, my friend and anthropologist Dr. Virginia Nazarea created an accepted technique that

As a young man, I was always more interested in growing and drawing things than I was about school. These interests still dominate my life. *Photo from author's collection.*

tries to maintain the links between crop diversity and cultural memory. In Virginia's book *Cultural Memory and Biodiversity* (2006), she sets out guidelines for documenting societal information about food crops from the people who have traditionally raised and used them. This protocol recommends recording everything from farming practices, names, dates, and locations of the people who grew the crops, along with stories, recipes, cooking techniques, and even songs. Much of her work demonstrates that traditional farming methods are often as important as more "advanced" methods used in growing certain varieties. Her work has been a tremendous influence on the way I think about seeds.

As much as I have always loved seeds and growing plants, I also love drawing things, especially things I grow myself. A literal example would be pulling up those Purple Top Turnips from my grandfather's farm and drawing a still life of them before eating them. It's just something I've done since I was a child.

Art and farming have always gone hand in hand for me. I've never been able to separate them. I would say that in my early years of grammar school, when I should've been paying attention to my lessons, I was looking out the window, daydreaming and drawing pictures. It took quite a bit of effort to get me through school and out into the world. When I graduated from high school, I went to the Ringling School of Art and Design in Sarasota, Florida, where I studied fine arts, landscape painting, and portrait drawing. When I graduated from there in 1966, I went on to the School of the Museum of Fine Arts in Boston. Inspired by my early fascination with the engravings in the *Maule Seed Cata-*

Starting in 1966, I studied at the School of the Museum of Fine Arts in Boston, and I returned there as an instructor in 1974. *Photo from author's collection.*

logue I had found in Knoxville, I focused on graphic arts, primarily printmaking, etching, and lithography.

In my youth, I struggled to make a living, and while I did all kinds of things to make ends meet, my passions have always been the same. After I finished graduate school, I spent two years in Europe, mostly in Austria and the Netherlands, where I worked in vineyards and with the sugarbeet harvest, and I continued painting and drawing. I remember some advice from an old man I met in Austria. He said, "Son, do something you love and you'll never work a day in your life." I really took that to heart. I picked the two poorest vocations out there: art and farming. It's hard to make a living in these occupations, but I wouldn't trade either one. I returned to the states in 1971 and worked as an artist in Sarasota, Florida, and then in 1974, I returned to Boston, as an artist and instructor at the School of the Museum of Fine Arts, where I taught for five years. Even in Boston, I did some farming on property owned by friends of mine. In 1978, while still living in Boston, I bought my first farm plot back in Newport, Tennessee, about forty-five miles from Knoxville. In 1985, I bought a second, adjacent tract of land that made up approximately two hundred acres. At that point, I was ready to leave the city and get home to my farming roots.

To me, art and farming both involve the creative process. Farming is about as creative as you can get. Some farmers may not necessarily think of it that way, but I believe there is a great art to planting seeds and nurturing them to their full potential. A crop cycle is a work in progress, just like a painting, throughout the whole season. My passion begins with a love for seeds and for what comes up and out of the soil. But then I like to record and preserve the things I grow through art, mainly drawing or painting. That's just another way of conserving them, as a still life or landscape painting. I

I love painting the landscapes I encounter in my travels.

One of the best ways to preserve seeds is to make sure they get into the hands of enthusiastic and responsible growers, like my good friend Louisiana chef John Folse. Years ago, I gave him a box of seeds I collected in Louisiana. Today, he grows them at his property, White Oak Plantation, where he serves them at his award-winning restaurants.

always admired and studied the works of the old Dutch masters, particularly the way they used light to capture fleeting moments in time and in nature. The natural world provides incredible inspiration for me artistically.

Seeds themselves are beautiful works of art. I love everything about them, the way they feel in my hands, their complexity of color, shape, and seed-coat patterns. To me, they look like little gems or polished stones. I love to display jars of them on my kitchen shelves, because they are so pretty to look at. I think of saving seeds as preserving nature's masterpieces, like we safeguard humanity's paintings and sculptures in museums. What if we said, "We've got the modern artists so we don't need

those Rembrandts or Vermeers anymore," and we just let the older works decay with time? If we lost our Old Masters and all we had left was the latest modern art, we would have pretty blank, empty museums. If we lose our heirloom seeds and farming heritage, we not only expose ourselves to the risk of global famine, but, as with paintings, we lose our connection to our cultural past.

Many friends have told me I was born in the wrong century. I have always gravitated toward people from an earlier time, and I have tremendous respect and fascination for the stories of how they lived off the land. Ever since I was a teenager, I would seek out old-timers and mine their experiences and memories for what many might consider the most mundane details of their lives. Agricultural practices have changed so much since the late 1880s that they are almost unrecognizable today. One of my oldest friends, Mr. Arlie O'Bryant, who was born in 1920 in the Thomas community of Washington Parish, Louisiana, always says, "The difference between farmin' today and farmin' when I was coming up is the same as going out th' gate backards." His expression has always resonated with me. Farm practices of the past are so completely dissonant from those of today. They represent two completely different worlds.

Homer Graves also lamented that loss. "By the 1960s it was all gone," Mr. Homer would say. "All the old methods of farming and way of life was pretty much run out by then." My friend Mr. Seldon Lang once boasted, "We were organic farmers and didn't know it." Before the use of mechanization, chemical fertilizers, and genetic crop modification, farmers possessed a wealth of practical and cultural knowledge amassed over generations. They knew everything, from how and when to grow certain crops, how to maintain the health of their soil and protect against pests and disease, and how to feed their families. They mostly passed this information down by working alongside their children, demonstrating practices and techniques through their daily activities on the farm and in the home. They certainly never wrote any of it down or saw any need to. As modernity spurred changes in agriculture and younger generations began leaving rural communities, there was suddenly no way to transmit this knowledge and no real use for it. To me it was precious, if only I could capture it all.

When an old person passes away, it's like burning down a library. As I saw these older folks dying off, I knew their accumulated farming knowledge was at risk. Starting in the early 1980s, I took every opportunity I could to work in the fields and kitchens with these

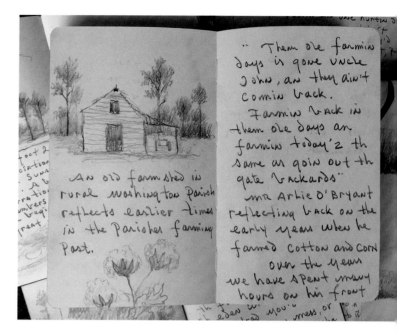

Sketch of a farm shed in rural Washington Parish, together with my friend Arlie O'Bryant's recollections of the old farming days.

men and women and to soak up their life's knowledge and record as much of it as I could while they were still around. I carried notebooks wherever I went to jot down what I observed and what I learned. My only regret is that I did not begin fifty years earlier, when there were so many more old-timers to talk to.

In the Smoky Mountains, where I live, I am constantly reminded of how our ancestors farmed those rugged hills, scratching their lives out of the ground without modern conveniences, and how they must have been a lot tougher than I am. As I plow the hillsides and work in the gardens and fields, I know I am following in their footsteps, planting the same crops, the same seeds. It's a cycle of life I am compelled to carry forward. I don't mean to exaggerate the importance of the material I have recorded in my notebooks. Much of it is in the form of rambling notes from conversations I've had and random sketches, and it was not intended to be read by anyone but me. But I do know how educational and rewarding it has been for me to write it all down, and how much this information has enhanced my life and informed my work. My hope is that I can demonstrate its value to others who are interested in preserving what is left of this farming heritage—a culture I fear much of America would allow to pass into oblivion without even acknowledging or mourning its loss. This book is my attempt to pass that trove of information along.

Many people have asked me over the years why I didn't think to use an audiotape machine or other device to record these stories. For one thing, technology and I are not suited for one another. And what would I do with hundreds of hours of audio recordings anyway? I am a man of the tactile world and always have been. I like the feel of pencil and paper. I like the look of notes and drawings, and I cherish the opportunity to pick up one of my journals, thumb through the pages I've created, and have them spark memories of cherished friendships. The idea of putting a recording device in front of someone I was visiting with would have been unnaturally intrusive and would have cut at the intimacy of plain old-fashioned conversation with periods of silent companionship and porch sitting. Many of my interview subjects and their acquaintances were already genuinely perplexed by my intense interest in their lives and my meticulous note-taking. Mr. Arlie O'Bryant often introduces me to his friends and neighbors with, "This is John, that one that writes all them books on me."

My early inspiration for documenting this material came to me in 1973 during a fortuitous visit I made to a small farming community in southeast Louisiana. I was between jobs and had returned to Sarasota, Florida, working as an artist, when I met Jennifer Vise, an

An October 2014 journal entry:

On the subject of recording stories during my visits to Washington Parish—the years when I spent time with Mr. Seldon Lang and later with Homer Graves and Mr. Arlie O'Bryant, a great deal of my time with them was spent in silence, as I listened to their stories, tales, and history lessons from the parish. I never wanted to miss a second's opportunity to record the treasures of their knowledge and experiences. I felt then as I do now, they were the masters and I was the student. Although I am now in my seventies I still consider myself as a young man who is "wet behind the ears" when I am in their presence.

For decades I have worked alongside my Washington Parish friends, especially Calvin and Rose Vise, who opened a door to a world that has enhanced my life in ways I can never fully express. *Photo from author's collection.*

I met Jennifer Vise while working as an artist in Sarasota, Florida, in the early 1970s. Jennifer introduced me to her family, and my life has never been the same. *Photo from author's collection.*

aspiring student studying at my alma mater, the Ringling School of Art and Design. Jennifer invited me home to visit her family in Louisiana. The Vise family had a small farm where they grew corn, beans, and a delicious array of fruits and vegetables year-round for their own use. They had nine children to feed, so eating out of the garden was a natural part of their lives. The rest, I would say, is history. I immediately fell in love with this place and I have returned to visit her family nearly every year since.

Washington Parish, Louisiana, is located in what is described as the toe of the state's "boot." It borders the state of Mississippi on two sides, is about ninety miles due north of New Orleans, and is on the way to nowhere. Washington Parish's location is a mystery even to most Louisianans. To get there, you have to be headed for Washington Parish, because it can only be reached on small, winding state roads. When I first visited these piney woods in the early 1970s, I had no idea what to expect. I had always heard stories of the state of Louisiana, which to me was the Creole and jazz culture in New Orleans, and the more rustic Cajun way of life in Acadiana—the name given to the parishes in the southern central and western part of the state. Imagine my surprise when I discovered Washington Parish, an odd combination of pine forests, hill country, remote cypress swamps, and a hardscrabble farm culture with humble but rich foodways.

I have vivid recollections of my first visit to Washington Parish and the morning I arrived in the small town of Franklinton by bus. The scene was shrouded in a misty fog that created a hazy impression of the surrounding swampland. I still can't explain my attraction to this place other than to say I felt an immediate, spiritual connection with the landscape and its residents. Everywhere you go you see large garden plots and small farms dotting the landscape. There is an intense bond between family and neighbors, and visitors are warmly embraced. People still gather to shell peas, put up corn, pick and can fruits and vegetables, and smoke meat. They also save and share seeds, although less than they used to. They remain tethered to the land and to each other more than in most places where I have spent time.

I was particularly taken by the people I met in Washington Parish, especially the older generation. I was intrigued by their expressive way of speaking, their masterful storytelling and colorful tales, and the traditional methods they used to prepare and share the food they harvested from the soil. Most profound was their overwhelming expressions of hospitality and generosity, which left me with a deep feeling of belonging. Regardless of age, color, or background, people there still address each other by first name, but honor each other with a preface of Mister or Miss, or Aunt or Uncle, even when they are not related. I've never met a stranger there. Even though Jennifer Vise and I eventually took different paths in life, the Vise family continued to open their doors to me as though I were one of their own.

As in most contemporary rural communities, the tug of modernity is strong in Washington Parish. In the past five decades the area has seen fewer people farming commercially. But even though young people are moving away, the traditions and culture of an earlier time remain relatively intact. This is what piqued my interest in writing and interviewing the older people I met, and in documenting everything I could. I think we all stumble on these rare opportunities in life. There are places you visit or people you meet, and you think to yourself, "I'm supposed to be here. I'm supposed to know these people. I'm supposed to be a part of this place. There's a story here and I'm supposed to be part of it, or at least record it." Never am I without pen and paper when I visit Washington Parish.

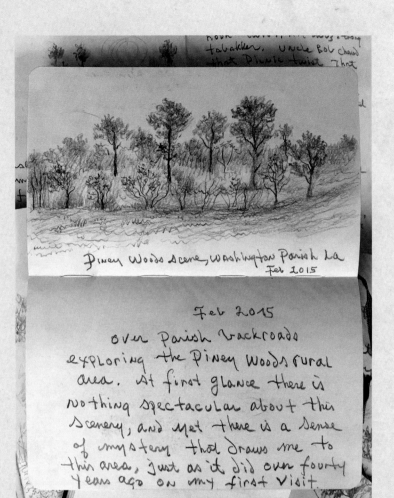

Piney Woods Scene, Washington Parish La
Feb 2015

Feb 2015
over Parish backroads
exploring the Piney Woods rural
area. At first glance there is
nothing spectacular about this
scenery, and yet there is a sense
of mystery that draws me to
this area, just as it did over fourty
years ago on my first visit.

A February 2015 illustration of the
piney woods landscape in and around
Franklinton, Louisiana, together with
my journal entry.

ABOUT WASHINGTON PARISH

The piney woods region of Louisiana has always prided itself on its fierce spirit of independence. This area of Louisiana lies north of Lake Pontchartrain, east of the Mississippi River and Bayou Manchac, and south of what is now the Mississippi state border. In colonial days, it fell within the Spanish territory of West Florida. Following the Louisiana Purchase in 1803, the United States asserted that this land was part of the transfer. Even though Spain disputed this claim, Anglo-American settlers flooded into the area. In 1810, Anglo settlers and British sympathizers who resented Spanish influence in the region mounted what was known as "The West Florida Rebellion" and established an independent Republic of West Florida. Leaders of the new republic wanted to remain independent of both the United States and neighboring Spanish territory. Only after a threat of force did the rebellious republic acquiesce to annexation by the United States. Today the heart of Louisiana's piney woods culture lies in the "Florida Parishes" of Washington, Tangipahoa, and St. Tammany. This region is known for gently rolling hills, small towns, farmsteads, and a rich growth of pine forests.

The new settlers in this region started small farms and raised subsistence crops, along with free-range cattle and hogs. Unlike most of Catholic Louisiana's French, Acadian, and Spanish residents, many early Florida Parish homesteaders were of Protestant Scotch-Irish descent and came from the Carolinas and Virginia, or by way of Georgia and Mississippi. They moved into areas inhabited by native peoples, mostly Choctaw, whose influence is evident through enduring names given to area rivers and creeks: Tchefuncte River, Bogue Chitto River, Natalbany River, Tangipahoa River, Bogue Falaya River, Bogue Lusa River, and Chappapeela Creek. While cotton plantations worked by enslaved Africans dominated much of the area leading up to the Civil War, after the war, white yeoman farmers who tended to reject most forms of governance emerged as the dominant ruling force, leading to a brutal state of lawlessness. As historian Samuel Hyde writes in his book *Pistols and Politics,* during the remainder of the nineteenth century this part of Louisiana had one of the highest murder rates in America.

The post–Civil War era saw freed slaves who had worked the sprawling plantations compete with poor white sharecroppers and subsistence farmers in scratching livelihoods out of the ground. White families, however poor, still held sway over former slaves and the free people of color who had moved north into the area out of New Orleans. They were all joined by migrant timber workers and European immigrants from Sicily and Hungary who were drawn to St. Tammany and Tangipahoa's cheap land and jobs.

Although the Illinois Central Railroad, built in the 1850s, stretched through what is now Tangipahoa Parish, there were no rail connections into the sparsely populated parishes until the early 1900s, when national lumber companies began clear-cutting the forests and connecting isolated communities with railroad lines. The Great Southern Lumber Company built a lumber town in Bogalusa, and by the 1930s, graded roads for automobiles began to reach into even the most rural areas. Still, the majority of farm families remained without electricity until the 1940s. Despite drastic cultural changes that would take place in most of rural America in the nineteenth and twentieth centuries, the people of this region re- mained virtually cut off from the outside world until World War II.

Many Washington Parish residents recall times in the not-so-distant past of cooking on woodburning stoves and attending school in a one-room school- house when they weren't out working in the fields. Despite inevitable modern- ization in American farming, however, Washington Parish remains deeply rooted in its humble past. As anthropologist Joy Jackson describes it in her book *Folklife in the Florida Parishes,* "Still, they retain the values and characteristics of that back- woods culture: a Spartan acceptance of adversity, a strong fundamentalist Prot- estant religious faith (usually Baptist or Methodist), a regimen of hard work from sunup to sundown, a passionate love and pride in their own land and the communal landscape in general, and finally a deep attachment to family—to its ancestral founders buried in family cemeteries located on the old homeplace, and to all the extended living members, whom they hope to see and visit with at the next family reunion."

I wrote about the intimate relationship between farmers and the land in a 2005 entry, "The Hands of Time":

I always found that a farmer's hands were like a roadmap of the life that he had lived. These were hands that were large from years of hoeing corn and cotton, and holding plow handles, plowing from sunup to sundown. Those were calloused hands that were as tough as whet leather, with deep tan lines from a life spent under the sun. Those hands read like a book, filled with their own story of a long life of hard work on the farm.

Arlie O'Bryant worked the land he lived on for more than half a century, and that toil is evident in the lines of his hands and face.

Like people and seeds, landscapes have stories to tell. You don't have to observe a particular place for long before those tales reveal themselves. Driving through the backwoods of Washington Parish, I sense something entirely unique and beautiful. Perhaps it is the year-round growing season, the wide variety of crops, and the rich tradition of seed saving and growing things. Or maybe it's a particular slant of light at certain times of day and the mood it creates, or the way light diffuses through the cypress trees or the pines as they cast shadows across the fields at sunset. I am drawn to the visual appearance of this region's enchanting landscape as much as I am to the people who call it home. It brings me peace and a comforting feeling of continuity and timelessness. And just like preserving seeds and stories from the people I meet, I am compelled to capture ephemeral moments in the Washington Parish landscape through art.

To me, the contours of the landscapes mirror the lines on the faces of the people that live on this land. You meet someone like my good friend Mr. Arlie O'Bryant. You notice the wrinkles in his old hands, and the deep ridges in his face, the lines of time. You see his leathered, weather-beaten skin, baked in the sun over years and years. You see character and tradition in the landscape, too. You can tell those farmers grew up on that land, that they dropped their sweat in that sandy soil, plowing mules, tending crops, and cultivating their lives. The land and the people are inextricably woven together in a poetic form of dignity.

During that first fateful visit to Washington Parish in 1973 with my art-school friend Jennifer, I met her parents, Calvin and Rose Vise. I was struck by the couple's youth and vitality. They had married as teenagers, and Jennifer was the oldest of their nine children. In the early 1950s, during Louisiana's early oil and gas boom, Mr. Calvin had moved the family from Alabama to Louisiana

Working in the field with my friend Calvin Vise. *Photo from author's collection.*

for a job with the Southern Natural Gas Company. Even though Mr. Calvin and his family were fully embraced as fixtures in the community and had lived there for twenty years when I first met him, he was still known locally as a "come here man," meaning his people were not among the original settlers in the region.

As a young man, Mr. Calvin worked offshore in the Gulf of Mexico and on pipelines, and he would be away weeks at a time. Like many families in Washington Parish, he always had a large garden, but because he was gone so much, he didn't have time to keep it up. I remember he grew all manner of things, including okra, corn, tomatoes, peppers, and squash. Even on that first visit to Louisiana, I was so intrigued by his garden that I made it my business to go out there every day and clear it of weeds and to spit-shine that land down to nice clean earth. I loved the opportunity to say I had my hands in Washington Parish soil, and that I worked with Washington Parish plants. Over the course of our forty-year friendship, Calvin Vise and I spent time together working in his fields, and cutting and splitting endless stacks of wood for his stoves. We tended and planted numerous fall and winter gardens in kale, cabbage, collards, mustard greens, turnips, winter onions, and garlic. I learned much of what I know about gardening from him. I also enjoyed many delicious meals at the Vise family table.

Through the Vises, I met many more friends and community members, who have all been generously welcoming. While I know that I, too, will always be considered a "come here man" in this part of the world, nothing makes me feel more at home than having a dear old friend call me "Uncle John," signifying my acceptance. These people have filled my life with treasured memories. There have been celebratory family gatherings around bountiful tables and many casual conversations around warm, slow-burning fires, or in rockers on the front porch. I'll always remember the camaraderie of laboring side by side in the fields, sharing rich stories of tough but well-lived lives. I have been blessed to develop deep friendships with many of these folks. From them, I have recorded, literally, volumes of notes, tales, and sketches from Washington Parish in my journals.

Another significant inspiration for writing these stories down was Mr. Seldon Lang, whom I met in the early 1980s. No visit to Louisiana was ever complete without a special trip to his home. While he was not a worldly man, Mr. Seldon was extremely well-read and learned, and a bit of a tinkerer. He was famous for being one of Washington Parish's great old storytellers, as was his father, Albert Lang, who had homesteaded the family farm in the early 1900s. Many of Mr. Seldon's spellbinding tales had come from his father's time in the 1800s and 1900s. Born in 1912, Mr. Seldon farmed a large plot of land year-round. Aside from his storytelling, he was well-known for growing and giving away his prized tomatoes at a roadside stand at the edge of his property, just so people would stop and talk to him. My visits with the Langs became a central part of my annual travel ritual.

Journal entries from the early days of my Louisiana visits express the excitement and anticipation I feel leading up to my visits, like these notes and reflections from my May 1989 visit:

Ritual of getting ready for Louisiana trip, wrapping tomatoes and pepper plants, loading truck beforehand, taking tools. Off for Louisiana at 3 a.m. . . . Low drifting misty clouds reveal a quarter moon. Misty hills at sunrise. Unique lighting on the vivid late spring greens. Drifting clouds of a hazy nature with filtered light over highlighted areas of the landscape down through Alabama

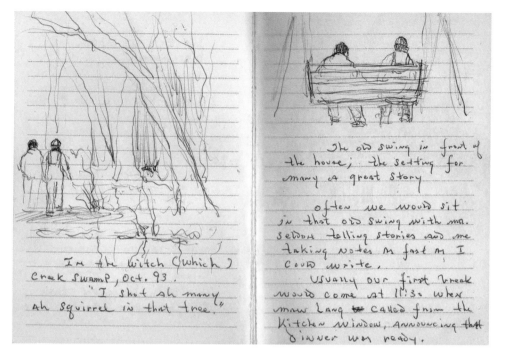

Many of my favorite visits with Mr. Seldon involved long walks on his property and hours sitting on his porch swing, where he would tell me stories about his childhood.

and into Mississippi. Shadows and light of a strong nature across the pine country all along the way. There is a strong sense of friends and family awaiting my arrival. I imagine what Paw Lang must be doing now. Each scene is so familiar as though I just saw it only moments ago. Highway 59 South has come to be my friend and companion along the way. Each scene is like an old friend revisited. That last stretch of road, Mississippi into Louisiana and dinner at Miss Iva Stewart's restaurant is waiting. When I sit down to rest and see the familiar faces, it is as if I never left. Much rain this season, beautiful shadows of green compared to last year's parched browns. Beautiful rolling landscapes along Washington Parish roads to Franklinton. As I drove up I could see Paw in overalls at the end of the road. Being back again and welcomed by friends and place with a "Welcome home weary traveler." Old friends glad to see each other. We

took our usual crop tour. All looking well. I gave Paw two bags of Red Ripper Cowpeas, a return gift from the stock he gave me in '87. We rode out to see his son Bill and his wife Daisy over in Sheraton. Good tomatoes and peas, bean crops. We sat under the shade of the dogwoods for a welcome reunion discourse after the long journey. I head down to the Vises', so much like home and family welcome. Mr. Calvin has done so much with the new land, cleaning, farming, fencing. Pound cake, baked for my homecoming dinner. Tomato and zucchinis, crowders with snap beans and stuffed peppers, cornbread, head of lettuce from my gardens. After dinner Mrs. Rose and I planted the tomato and peppers that I brought down. A beautiful working scene at dusk all three working to complete the planting. Such a sense of fulfillment and well-being.

SEASONS OF LIFE | 33

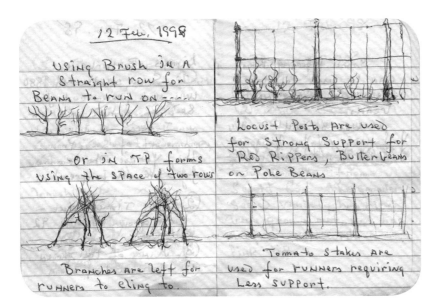

My friends shared invaluable growing tips with me that I still use today.

Below: Visiting with Seldon Lang.

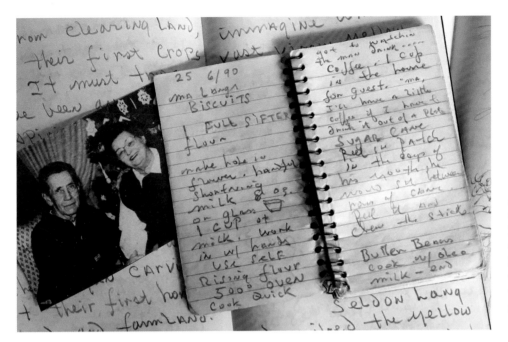

A June, 26, 1990, journal entry describes morning biscuit-making lessons with Maw Lang. What I learned in watching her was that she did not cook with recipes or use precise measurements.

For years, I spent countless hours sitting on Mr. Seldon's porch swing listening to stories and learning from him. While he loved conversation, he was not too happy when I took notes during our visits. So that I would not leave any details out of my notebooks, I often excused myself after an hour or so of storytelling and hid behind the barn, where I feverishly wrote down everything I could remember, before resuming my seat beside him. These listening sessions served as a pleasant interlude between mouthwatering family meals that drew friends and relatives from near and far. Mr. Seldon's wife, Beulah Mae, was a highly regarded cook, but also a beloved and revered midwife. Long after her retirement, her presence was still requested at births all over the parish. Some of my fondest memories of visits to their home include cooking lessons in Maw Lang's kitchen, and fieldwork at Paw Lang's side.

While this Deep South region is not known for the rich fertile soil found in the Mississippi River Delta to the west, it has an enviable growing season and a good mixture of sand, silt, and clay well suited for row cropping. As I spent more and more time with Mr. Seldon and Mr. Calvin, and as I met more and more people in this community, I discovered that I had never seen or heard about many of the crop varieties they grew. For example, in my early visits in the 1970s, I had no idea there were so many different varieties of field peas and butter beans, which are in the lima family. Butter beans are occasionally grown in Tennessee, but not to the extent that you find them in the Deep South. In my ongoing quest for seeds, I came across numerous, unusual varieties of beans, peas, okra, corn, and squash in this area. These old varieties and their histories fascinated me, and I immediately began to save and document them, too.

One of my most prized possessions is a variety of field pea Mr. Seldon gave me in 1987. As I was leaving

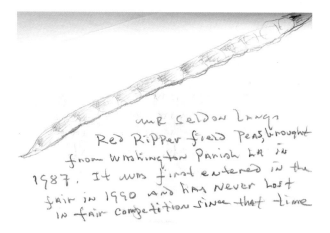

An October 2, 2005, journal entry describes Mr. Seldon's prized peas, a gift he gave me.

Red Ripper cowpeas are not only beautiful—they are delicious.

to return to Tennessee, he handed me a small paper bag of medium-sized red peas, which he labeled "Red Ripper Pea for the Corn Patch." For years, he had talked about this variety and how his family had grown them. The Red Ripper Cowpea dates back to 1830 and was common in Texas and Louisiana. This particular strain is larger than most, and I was thrilled to receive the gift.

As soon as I returned home, I jumped out of the truck. I didn't take my suitcase inside or anything. I went straight down to the corn patch and dug my "new" peas in around my corn. I was so anxious to have them grow, I took a watering can and carefully dampened each hill of peas and watered them often so they would come up extra fast. That first crop was wonderful, and I continued to grow and save the seed stock that came from that planting. In future years, I would work with Mr. Seldon and plant these same peas back in his gardens. His variety of Red Ripper Cowpea is the best field pea I ever tasted. When I share those peas today with other people, I always think of him and the time we spent together in the fields.

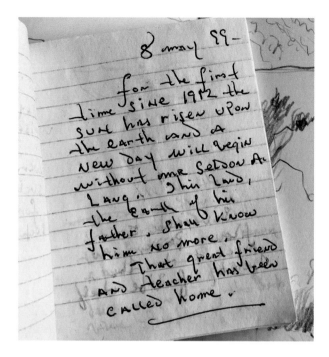

A May 8, 1999, journal entry about the passing of my good friend Seldon Lang, a devastating loss.

After those long workdays, we'd sit on wooden crates in the shade of tall pines, where the late afternoon sun illuminated the spot where the original Lang home once stood. In one 1989 journal entry, I reflected on one such day that is etched in my mind: "This was one of those precious times in life, spent with one whom I admire, look up to as a hero of sorts and certainly one of my greatest teachers, I feel I am in the presence of one of life's special gurus." After hours of listening to Mr. Seldon's stories and advice, I wrote that in this man's presence I saw an uncommon strength and guidance to be drawn upon:

> I am fortunate enough to be blessed with these
> character-building moments, a short time in passing

which will last a lifetime. I was to think of Paw Lang often during the long trip home on the back Louisiana roads and the country roads up through Mississippi. As the world passes outside my window, I recall a partial story, some advice or time spent, hoeing a long hot row together, or the wonderful times at Maw Lang's dinner table. I think of how much life has been enhanced and how much richer I am for having been privileged enough to have known and shared this man's knowledge and wisdom. The pines grow rich with the colors of the late afternoon and the stories of the old days told.

For more than a decade, Mr. Seldon was one of my biggest resources of Washington Parish stories and farming information. On one of our early visits I asked him to tell me everything he knew about growing field peas. He must have thought I was an unusual kind of fellow because he looked at me with the strangest expression and barked, "Good Lord, son, a blind man can make peas." I'm sure he had never given the process much thought or considered the value I placed on that type of information, but to this day I consider his expertise and the time that I spent working alongside him one of the greatest gifts of my life. He taught me how to grow peas, onions, and sweet potatoes, as well as the secret to growing his prized tomatoes. He showed me how to repair antique farm tools and build outbuildings and fences on the land his father had homesteaded, and where his mother's house once stood, next to the pear tree she had planted. He told me music ran through his family and that he and his father learned to play the fiddle, and how he made his own instruments. His simple wisdom and decency became a tremendous influence on my life and work. His loss in 1999 devastated me; he had become a mentor and a real father figure.

Mr. Seldon's passing, coupled with the loss of many

other friends in Washington Parish in the late 1980s, made me realize that time was running out and that I needed to get serious about recording the information they had shared. With this awareness, I developed the need to hang on for an extra couple of minutes each time I visited. Suppose this was the last chance I had to talk with them. Maybe there was one more valuable story, a bit of information, or a saying to record before I left. These interviews, frantically but meticulously recorded, now fill the majority of my journal pages.

When I first started writing these journals, I wasn't very sophisticated. I just carried in my pocket these tiny spiral notebooks that you buy at the five-and-dime store. I used to jot down reflections on my trips, sparse notes from conversations with Mr. Seldon or Mr. Calvin, meals that I ate, the addresses of people who wanted to share seeds, "to do" lists, and sketches of things that caught my eye. Years later, I switched to larger leather-bound notebooks and kept writing and drawing. I am still not sure exactly how many notebooks I have written altogether. From my Washington Parish travels alone, I have one big box full of twenty or so small spiral notebooks. I know I have more than eighty of the larger journals and, every once in a while, I still come across some that were tucked away and forgotten. With each new visit, I usually add three or four more new books, so the collection continues to grow.

I write down anything at all that comes up during conversations, covering a wide range of topics over the expanse of only a few pages. There are family histories and family stories. There are hunting tales, local folk tales, and ghost stories. I've recorded figures of speech and sayings that struck me as unique or humorous. Many old-timers told me about what their days were like as kids on the farm and the types of chores they did. I've documented minute details about specific heirloom

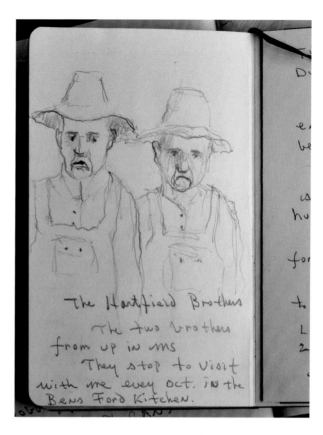

An October 13, 2008, sketch of two brothers who always visit the kitchen where I volunteer in the Mile Branch Settlement during the Washington Parish Free Fair.

crops, illustrated descriptions of seeds and plants, tips on how and when to plant, and I've diagrammed techniques and old farming implements used to grow the plants.

Then there is the food. Everything I've ever eaten in Washington Parish, I've written down in some form. I've never found anywhere I like better cuisine-wise than Louisiana in general, and this region in particular. Meals still center on things that come out of the field. It is a tradition born out of necessity during a time when

Back in them years hso
All ther Land wus in
cotton an corn. you didn

see fields all growd up like they are
today, them fields wus yor
livin, you didn have no other
way ah makin it. when you made
ah dollar, that dollar wus hard
come by, an they wadn many
dollars 't be made. we wus all
time huntin up some way 't
make ah extra dollar. Homer Graves

A February 21, 2015, sketch of the fields of corn and cotton that once dominated the Washington Parish landscape, together with Homer Graves's recollections of earlier times.

a family had to be self-sufficient in putting food on the table, making land essential for survival. That was the dominant way of life in this area until after World War II, when farmland was converted into more profitable pastureland or timberland and eventual mechanization and large-scale agribusiness began choking out smaller family farms.

Even today, eating serves almost as a ritual, where people come together at the family table over familiar dishes to break bread. "Dinner" is the largest meal, served at midday. "Supper" is more of an evening snack. I am grateful to have been welcomed at these family meals and have been privileged to look over the shoulders of these seasoned cooks as they prepare them, writing down everything I can between samplings. Many of these women recalled how their mothers and grandmothers prepared meals on open hearths and on wood-burning cook stoves, and how much better food cooked that way tasted. They can describe with great detail the difference in flavors between varieties of cornmeal and field peas, and wood for smoking meat. I have worked to train my palate to recognize the subtleties they describe, and they are right.

Until more modern farming practices were adopted, men and boys generally did the hard field and physical labor. Younger children did whatever work they were told to do, and women tended homes and gardens and prepared meals from what their families had on hand. Like farming expertise, cooking know-how was not written down; it was part of an oral tradition that was passed from mother to daughter. Recipes didn't use measurements, but relied on learned techniques. I've done my best to capture many of these time-tested methods, complete with a "pinch" of this and a "handful" of that, and I've recorded memories about canning, storing, and preserving food and seed for future use. In this book, I share as many of these treasured family recipes and techniques for preparing food as I can.

Listening to my friends' stories, I find that many of our dearest memories from childhood are enhanced with nostalgia. As the years pass we often embellish the simplest recollections, elevating them to the highest levels of milestones that mark special events. One such memory was related to me by Mr. Seldon. After a fine dinner, as we sat on the old swing near the house, he recalled one such experience from when he was just ten years old. His neighbors Uncle Frank and Aunt Mag Douglas had come to spend a Sunday afternoon with the Lang family. This was a rare opportunity to visit and catch up on social news, something that many were deprived of due to their isolation.

As the day's visit drew to a close and the late afternoon shadows began to stretch across the yard, Aunt Mag asked the young Mr. Seldon if he would like to come home with them and stay the night. For a young boy, this was a momentous opportunity. It was no more than four or five miles as the crow flies to Aunt Mag's. But to little Seldon, whose concepts of the world stretched little farther than the boundaries of their farm, a trip to his neighbor's house was as special as a journey to some far-off exotic place.

Bidding farewell to his family, young Seldon climbed aboard the wagon with Aunt Mag and Frank. Frank drove the team of mules, and Aunt Mag sat beside him on the buckboard. There was not a great distance to travel, but the going was slow on the old rutty dirt road. By the time they crossed the Bogalusa Creek, the sun was setting fast.

On that ride, Frank talked of running fishing lines along the creek banks that night, and Aunt Mag reflected on her garden produce in need of canning in the next day or so. To say the very least, all three travelers had worked up a hearty appetite, and the question "What's for supper?" had come up, several times. As the last leg of the journey approached, Frank turned the mules off the road and the travelers headed down the sandy lane towards home.

By the time they reached the house, twilight's last glow was quickly fading, and the Sunday dinner wearing rather thin. As Seldon climbed down from the wagon, Aunt Mag told Frank to go to the corn patch and pull some "roasting ears" for fried corn. Once inside the kitchen, Aunt Mag lit a coal-oil lamp, which illuminated the board-and-batten walls and the room's furnishings in a soft warm light. The next order of business was starting a fire. From wood in the woodbox, she shaved off thin splinters of lightered wood kindling, and she threw it and split red oak into the stove's firebox. Lightered wood, also called fatwood, is the cured, sappy heartwood of pine trees that is highly prized for its burning qualities. Soon there was a hot fire, along with the sounds of a crackling, popping blaze and a strong draft rushing up the stovepipe. The smell of lightered and oak burning whetted the young Seldon's appetite.

Frank quickly returned with an armload of roasting ears and was put to work shucking the corn. He cut the kernels from the cobs and scraped the corn "milk" from those cobs. The corn and milk were fried with bacon grease in an old cast-iron skillet, a favorite way to cook it. Aunt Mag immediately busied herself with peeling and slicing potatoes for frying, and then set about making biscuits. This was well before the days of vegetable shortening as we know it, so instead of shortening, she used lard that she kept in a large tin storage tub. She brought the dough together with rich whole milk brought in from the springhouse. Soon the kitchen was filled with a marriage of heavenly aromas: fried corn simmering in the large black cast-iron skillet, potatoes in another skillet turning a beautiful golden-brown, and biscuits rising in the oven.

Mr. Seldon remembered his taste buds watering in anticipation; the aromas of the cooking food overwhelmed him. Aunt Mag pulled the hot, fluffy biscuits from the oven and set them on the table with bowls of fried corn and potatoes. As a sweetener for the biscuits, there was a tin of ribbon cane syrup. Mr. Seldon reminded me that "Hunger is its own best sauce." In his

Visiting with my friends Homer Graves (*center*), and
Calvin Vise (*right*).

young mind, as humble as this meal was, no gourmet banquet could have matched it in terms of pure satisfaction.

By the time supper was finished, the hour was late and it was past his bedtime. From his bed by the window, Mr. Seldon recalled watching the full moon rise higher in the sky, bathing the piney woods in an eerie silvery blue light. Perched on a dead limb of a nearby longleaf yellow pine was a great horned owl, a magnificent silhouetted statue against the background of the full moon. The scene faded as the dream world outside turned to the dream world within. In telling me this story seventy-two years later, Mr. Seldon clearly regarded it as one of his life's special memories.

Many of the special memories that I have recorded in my journals include sage bits of wisdom, an amusing figure of speech, or a unique way of referring to something familiar in everyday life. In my journals, I respectfully attempt to record vernacular expressions that people use, copying them down just as I hear them, because the way people talk is often more interesting to me than what they have to say. If I tried to translate these expressions into modern English, I fear the meaning and character would be lost, and I would feel I'd attempted to "correct" what was said to me. Often times, I write down a single word or a snippet of a story that I try to go back and complete later, if I remember. In my mind, it is essential to record everything. Some of my favorite sayings came from my friend Homer Graves. Mr. Homer had so many wonderful expressions that I began referring to them as "Homerisms," and I use them in conversation myself, practically every day.

My journals are also filled with drawings of things I see around Washington Parish. I'll sketch landscapes, trees, and fields, or components of plants and seeds. I draw architectural details of old farm buildings and

An October 29, 2010, journal entry with notes on conversations with Homer Graves.

barns, and picket fences and antique farm tools. I might come across an old saw or ax, hill sweep, harrow, or plow stock, tools for smoking meat, or an unusual milk jug, canning jar, or handmade basket. These items may not be unique in the historical sense. At one time they were common everywhere, but they're uncommon now because they're not used today.

One of the main things I've been able to document are varieties of vegetables and fruits once abundant in this region, or that were safeguarded and grown by

An October 13, 2010, journal entry about varieties of fruits and vegetables grown in Washington Parish.

generations of local families, but have since disappeared. In some instances, I have spent decades trying to locate certain varieties. I have had some degree of success. That gives me a living connection not only with the people I know and spend time with, but also with the seeds I grow and preserve. It's a living heritage that goes along with the information preserved in my journals. When I find rare varieties from Washington Parish, I take them back home to Tennessee, where I propagate them on my own land or at Blackberry Farm. I safeguard them with Seed Savers Exchange and return seed stock to Washington Parish, where I hope to see them reestablished with regional growers. I try to encourage the next generation to take on this responsibility. I am always looking for young people who are going to pass these seeds down and grow them and care for them and hopefully pass them on to others.

Finding old seed varieties takes a good deal of detective work and a dose of raw luck. Each time I hear about a variety unfamiliar to me, I record every detail from my source. I dig up as much information as I can about the family known to grow the seeds, and what the seeds and plants looked like, so I will know it if I come across it. My journals are full of sketches of a bean, of a plant, and of a field planted in a certain crop the way I imagine it would look. Sometimes people hear that I am looking for seeds and bring them to me. That is how I came across the greatest prize of all—the Unknown Pea of Washington Parish.

Since I first began talking to people in Washington Parish, I heard about the old pea commonly grown in this area, called the "Unknown Pea." All the old-timers told me about it and how it was common in every cornfield in the parish going back before the 1900s.

No one knew where the tiny pea came from, so that's the name that was attached to it: The Unknown Pea of Washington Parish. What an odd name—it doesn't sound real. I thought, my goodness, I need a sample of that pea to get it started and preserve it. I even searched through old agricultural extension records at Louisiana State University, where I found a 1917 record and photo documenting the Unknown Pea, and have since found it documented in other regional agricultural records.

As I continued my search, all the old-timers said the Unknown Pea was gone and that no one grew it anymore. It is unthinkable that it had gone from being the most common one grown to being assumed extinct. For thirty years I looked for that Unknown Pea. Finally, in 2010, my good friend Gus McGee brought me a small pill bottle full of them. He got it from a man named Dan Seal, who was eighty-seven years old at the time. Mr. Seal is since deceased. Now, Mr. Seal may or may not have been the last one to have the Unknown Pea of Washington Parish, but now that I have it, I certainly treat it as though I'm the only one that's got it. To safeguard it, I've since sent a sample to Seed Savers Exchange. I have also brought seed stock back to a number of people in Washington Parish. I hope to see the Unknown Pea reestablished as part of the history and heritage of this region. Best of all, Mrs. Letha was right about those peas being delicious. I think the Unknown Pea of Washington Parish is one of the best-tasting peas I've ever had.

———◦———

One of my greatest sources of information for my journals, as well as my favorite event of the year, is the annual Washington Parish Fair in Franklinton. My first time attending was 1988, and I have worked as a volunteer at the fair every year since. The Washington Parish

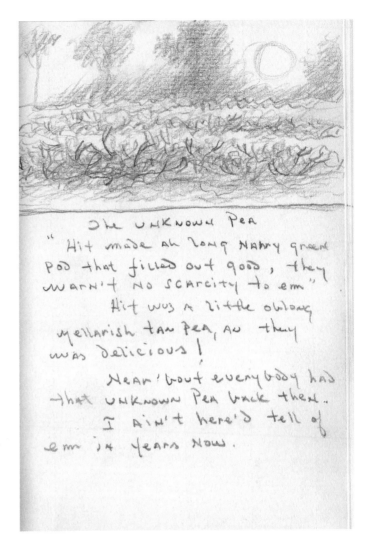

During a Louisiana visit in 2000, I sketched an imaginary scene of a field planted in the mysterious Unknown Pea and recorded a description told to me by a fellow fair volunteer, Mrs. Letha Toney. Her description of the pea helped me to locate it ten years later.

"A close call;
unknown pea, once
now found.

The unknown p-
...shington Parish havin-
...hich was once thoug-
...was found +o

THE UNKNOWN PEA

Since I first began talking to old-time farmers in Washington Parish, I had heard about the old pea that was once commonly grown in this area, and it was called the Unknown Pea. The old variety dates back as far as anyone could remember, and the first person to tell me about the variety was my old friend Mr. Seldon Lang, who remembered his father, Mr. Albert Lang, growing them down through the years. No one seemed to know where the peas came from originally or if they ever had an official name. In Washington Parish, they were simply known as the Unknown Pea. Miss Letha Toney described the peas for me as having a "fairly long pod that was solid green, an the peas wuz little oblong light tan colored. . . . Them little peas really made too! They warn't no scarcity to 'em, an them vines wuz hanging full." For well over twenty-five years I searched for this old variety and asked old-timers if they knew anyone who might still be growing them. I asked a number of people to be on the lookout for them. Down through the years, all efforts met with no leads or results, until the day when Mr. Gus Magee showed up with the bottle in his hand and said, "I've got something for you." And there they were, the long-lost Unknown Peas, found at last!

Gus had been searching for these peas for several years, and he had just about exhausted all possible sources when he finally talked to a Mr. Dan Seal, a farmer in Washington Parish who is in his late eighties. The Unknown Pea, once lost and now found! Finding this pea was a great discovery, but also very well illustrates just how perilously close so many

The Unknown Pea of Washington Parish.

of our old varieties are to being lost forever. Many have already been lost. Thankfully, others have been saved for future generations. The Unknown Pea was once commonly grown on farms all over the parish, and seed was saved by individual families down through the years. The first thought that comes to mind is how do we go from a time when almost everyone was growing these peas to a time when they were thought to be entirely lost? Having been a seed saver for many years I can answer most of my own questions concerning genetic erosion, changes in farming methods, the introduction of new crop varieties, large percentage of the community no longer farming, and the list goes on. The story of the Unknown Pea has a happy ending, but other varieties have not fared as well and are now gone forever. Now more than ever, it is urgent that we continue to search for old varieties that are in danger of being lost.

A list of field peas documented in agricultural journals at LSU's Hill Memorial Special Collections Library. From Louisiana Cooperative Extension service photographic album, 1906–79, page 37. Courtesy of Photographs Collection, Louisiana Cooperative Extension Service Records, A3003, Louisiana State University Archives, Baton Rouge.

The third week of every October is the Washington Parish Free Fair, billed as "the largest free fair in America."

Every year, I volunteer to cook the type of meals that farm families a century ago would have cooked on the woodstove in the Ben's Ford Kitchen. The kitchen is a historic log building that was moved to the Mile Branch Settlement in the 1980s. *Courtesy of the Varnado Museum.*

Fair is a community institution. The town literally shuts down and area schools let out for the week so children, families, churches, and community groups can attend. Founded in 1911, the Free Fair is the second oldest fair in Louisiana, and it regards itself as the largest free fair in the United States, meaning they don't charge admission, but there is plenty of good food and local crafts for sale. Franklinton is a small town with a population just short of 4,000; however, for four days during the fair every October, it turns into a big city of more than 125,000.

In its humble beginnings, the Washington Parish Fair started as an agricultural exposition and as a way to educate farmers about the latest farming techniques and new crop varieties. To this day, farm exhibits of livestock and homegrown vegetable and flower competitions are central to the fair, and people of all ages participate. Young people from all over the state compete in rodeo events and march in the parade. As it has for more than one hundred years, the fairway features carnival rides and games, with several stages packed with live music and local choral groups. Many people view the fair as a family reunion, and they often see friends and neighbors they haven't visited with since the previous year's fair.

At the center of the fairgrounds is the Mile Branch Settlement, where historic farmhouses and restored old buildings from the area have been moved to form a small village. The venue has several styles of homes, a post office, chapel, schoolhouse, meal house, general store, sawmill, and gristmill. The village is staffed by volunteer reenactors who perform demonstrations in traditional crafts and play old-time music. Volunteers fry cracklin', and they grind cornmeal and sugar cane, and make cane syrup, lye soap, quilts, and baskets.

Since 1989, I have volunteered in the Mile Branch Settlement doing cooking demonstrations on a wood-burning stove in a pioneer log cabin known as the Ben's Ford Kitchen. In 2016, organizers of the Washington Parish Fair surprised me at a special ceremony with a tremendous honor for my years of service. They gave me the symbolic keys to the city of Franklinton and a plaque with my name on it. I just love the fair, and it is part of a yearly ritual as central to my life as the inevitable changing of the seasons. Each year, I look forward to new friends, new experiences, new seeds, and new stories to document. Thousands of people have passed through that kitchen over the years, and I have done my best to record their stories. Year after year, unfortunately, the number of older faces I look forward to seeing each fall dwindles. But when I look ahead with anticipation for the promises of a new year, I also look backward with appreciation for those who have come before me. I hope to repay these steadfast folk by sharing the stories of the seasons of their lives. It is my fondest hope that through this book their memories will live on.

WINTER

Nothing makes you more conscious of the cyclical nature of life than farming. There is no start or finish, no beginning or end. In times past, the coming year's crop depended largely on the previous year's yield. Preparations made in fall and winter to save seeds, build soil, and prepare fields improved the chance of success in the coming spring. Even as days grew shorter, the work was never done. The land rested in winter, but farm families did not. During the nineteenth and early twentieth centuries, winter on farms like those in Washington Parish was a time to get ready for the spring planting season and to catch up on tasks there was no time for during the busy harvest.

At the start of the calendar year, in the dead of winter, Washington Parish's farm families lived off what they had around them. Mr. Seldon Lang would say, "If you didn't shoot it out of the tree, fish it out of the river, or dig it out of the ground, you didn't have it." Winter farm tables featured mostly cold-weather crops, such as leafy greens and root vegetables, along with fruits and vegetables people had dried, canned, shelled, or stored during the summer and autumn harvests. Winter gardens were full of collard greens, mustard greens, turnips, and rutabagas. There were multiplier onions, garlic, and winter cabbage. Fresh and preserved produce was flavored with smoked meats. They were accompanied by seasonal trapped or hunted birds and game, when available. Cornbread was served at every meal.

One of my favorite Washington Parish traditions was to help Mr. Calvin Vise plant his winter garden each year. I would stay long enough to see things barely come up, but I couldn't be around to see the full results of our work. In my journals, I recorded the garden plantings by drawing whole garden plots and layouts so that, when I returned home to Tennessee, I could look back and wonder how all those fall onions and shallots and garlic might be doing.

Greens have always been a staple in the American South's gardens and kitchens. Farmers in Washington Parish grow a number of varieties. Favorite turnip varieties in the region include Seven Top, Shogoin, Purple Top, White Globe, and Yellow Globe. When it comes to cooking turnips, most often both the tops and the roots are boiled or steamed and served separately or together.

An end-of-1999 journal entry describes plans for future gardens and the growing fall garden that I helped plant that year at Calvin Vise's house.

Mustard greens are the broad, peppery-tasting leafy greens of the mustard plant. The yellow seed of the mustard plant is what's ground with vinegar to make prepared mustard. Tender mustard greens can be eaten raw, but they are most often simmered in water or broth and flavored with seasoned meat. Local favorites include Florida Broadleaf, Southern Giant Curl, and a number of more recently introduced varieties.

Collard greens are in a class of their own. Closely related to cabbage, collards have a character and flavor unrelated to any other cooking green. These hearty, dark plants have a thick, heavy leaf and require a bit more cooking than mustard greens. The most popular varieties in Washington Parish include Morris Heading, Southern Champion, Georgia Creole, and Vates.

I have never found the methods of preparing any variety of greens to vary a great deal. In most cases, the greens are thoroughly washed to remove grit and insects.

Some cooks I know tear the leaves before cooking, while others roll the leaves up and slice them uniformly. Cooking greens calls for a good piece of smokehouse meat, which can be bacon, fatback, middlin' meat, streaked meat, or any other cut of smoked pork. Cooking times vary with individual cooks; some don't cook young, tender greens more than a few minutes, while others hold to the old traditional method of slow cooking any kind of greens down for hours.

A few tips I have learned from Washington Parish cooks over the years include browning seasoning meat in the pot before adding the greens, which imparts a richer flavor. A dash of dried hot red pepper flakes also goes well with greens. If greens are fully mature and have a bitter taste, a small amount of sugar or a splash of vinegar improves the flavor. As is the case with many old-time recipes, there are no set measurements for ingredients.

Greens of all kinds grace winter gardens across the South.

COLLARD GREENS

Makes 6 servings.

———◇———

This is the usual way I cook greens. It's not fancy, but it sure tastes good, especially when I serve it with cornbread to sop up the potlikker, the juice. This recipe works with any kind of southern greens.

2 large bunches fresh collard greens

10 slices smoked, thick-sliced bacon, finely chopped, or

    1 smoked ham hock or ¾ pound smoked meat, chopped

1 small onion, finely chopped (optional)

3 cloves garlic, finely chopped

2 quarts water

½ teaspoon red pepper flakes or fresh hot pepper

Cider vinegar and sugar, to taste

Ruthie Mae Graves's cornbread

    (optional); recipe on page 63

1. Cut out tough stems from collard leaves. Hand tear leaves into smaller pieces. (Some cooks stack flattened leaves, roll them up, and slice them.) Thoroughly and repeatedly wash collard greens until wash water is clear. Set aside.

2. In a large cast-iron Dutch oven set over low heat, cook bacon until fat is rendered. Remove bacon from pan and leave in 2 tablespoons fat. Sauté onion and garlic in the grease until slightly caramelized, when it's soft and brown.

3. Return bacon to the mixture. Add water, pepper, and torn leaves. Bring everything to a boil and reduce to a simmer. Cook uncovered 2 hours, or to desired degree of tenderness. Add additional water if necessary to keep the greens submerged.

4. When greens are cooked, add vinegar and sugar to your taste. Serve greens warm. If you like, put greens and potlikker in a small bowl and serve it with cornbread.

RUTHIE MAE GRAVES'S TURNIPS AND GREENS WITH CORNBREAD DUMPLINGS

Makes 4 servings. Courtesy of Homer Graves.

———◇———

I had never heard of cornmeal dumplings cooked in pot-likker until about twenty years ago, when I had them for supper one night at Homer Graves's house. Homer's wife, Ruthie Mae, would make them when I visited, and I will never forget how they would just melt in your mouth.

    2 tablespoons bacon grease
    ¾ pound smoked, seasoned pork
    1 small onion, finely chopped
    ½ teaspoon chopped fresh garlic
    2 ½ quarts water
    ½ teaspoon salt
    ½ teaspoon ground black pepper
    1 bunch turnip greens, with roots
    Cornbread Dumplings (recipe follows)

1. Add bacon grease to a medium-sized heavy pot and sauté smoked pork a few minutes. Add onion and garlic and sauté until soft. Add water, salt, and pepper. Bring to a boil; then lower to a simmer.
2. Chop off the turnip greens from their roots. Slice the stems out of the turnip leaves. Tear the leaves into bite-size pieces and wash them thoroughly several times to be sure they are free of sand and grit. Add the leaves to the simmering liquid. Cook 1 hour, uncovered, stirring often. Add additional liquid as needed to cover the greens.
3. Peel and slice or quarter the turnip roots. Add sliced turnips to the greens and continue cooking until tender, approximately 20 minutes.

4. Using a large spoon, pull greens to the side of the pot. Use a tablespoon to drop cornbread-dumpling batter into the bubbling liquid. Shake the pot gently to be sure each dumpling is completely covered in liquid; do not stir. Boil until dumplings are cooked through, about 10 minutes. Serve warm turnip greens, turnip roots, and the dumplings together in a bowl.

CORNBREAD DUMPLINGS

Makes 6 small dumplings.

———◇———

    1 cup fresh, finely ground cornmeal
    ½ cup self-rising flour
    ½ cup potlikker (juice from the greens)
    1 large egg
    ½ teaspoon salt

Mix all ingredients together to make a thick batter. Drop from a tablespoon into hot, boiling liquid and cook about 10 minutes.

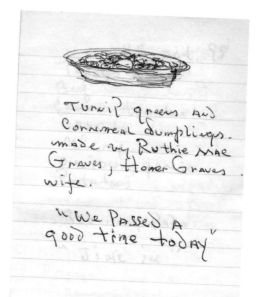

An October 22, 1998, sketch of Mrs. Ruthie Mae Graves's turnip greens and cornmeal dumplings.

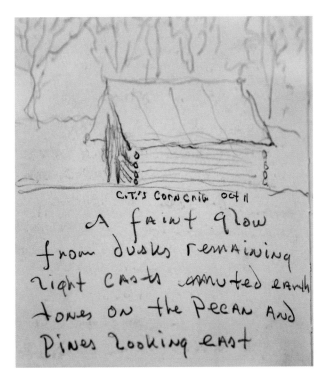

C.T.'s Corn Crib Oct 11

A faint glow
from dusks remaining
light casts connoted earth
tones on the Pecan and
Pines looking east

Left: An October 11, 1994, sketch details Calvin Vise's corn crib.

One of the most beautiful and tasty varieties of corn I have collected in Washington Parish is Pigott red-cob corn, which is used to make cornmeal.

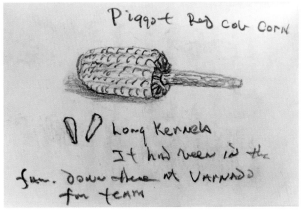

Pigott Red Cob Corn

Long Kernels
It had been in the
fam. down there at Varnado
for years

Corn was the staff of life on southern pioneer farms, serving as livestock feed and meal for bread. Dried corn was stored in outbuildings known as shuck cribs, where it was layered in between dried corn husks. Corn cribs were elevated to discourage pests and scavengers. Corn was carefully rationed to make sure there would be enough to get through the winter and until the next summer harvest. Mr. Seldon Lang told me more than once that if it was a sparse year for the corn harvest, the old-timers would say, "the wind's gonna blow through the corn cribs in March." In other words, by spring, you would be down to low rations.

In the early part of the twentieth century, certain varieties of corn were prized in Washington Parish for making the best cornmeal, especially yellow and white shoepeg corn. Shoepeg corn was named for its narrow-shaped kernels, which look like the small pegs once used in shoemaking. This variety has smaller, sweeter, more densely packed kernels than most corns. Today, shoepeg varieties are rarely grown and are difficult to find, as they've been replaced over time by modern hybrids with larger, more uniform kernels. In the past, farm families also cultivated and grew unique varieties that were descendants of corns grown by the area's native populations. Small-kernel popcorn and pencil cob corn, named for its narrow cob and long kernels, were also popular. One of the tastiest and most beautiful varieties I have come across is the Pigott family's red-cob corn, which has a beautiful orange gradient up to its red tips, and it makes a delicious cornmeal.

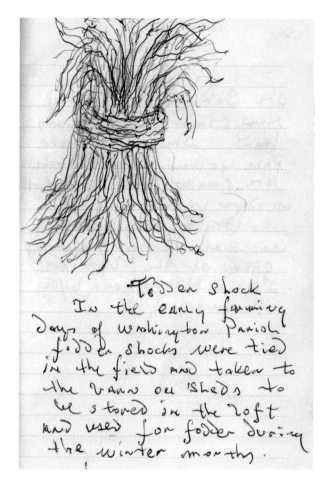

Everything grown in the field was used including corn stalks and husks, called "fodder shocks."

Around the turn of the century, cornmeal was a staple at every meal in many southern kitchens, especially in Washington Parish. Cornmeal is simply coarsely ground flour made from dried corn. It was used most often to make cornbread, which in lean times could be the main course. For variety, cooks sometimes added peppers or onions, or, best of all, the fried pork skin delicacy known as cracklings or "cracklins." Cornbread was generally served with freshly made butter, fruit preserves, or cane syrup.

Cooks were creative in their preparation of corn and cornmeal, making everything from cornmeal "mush" or grits, to cornpone, cornmeal dumplings, and hoecakes. "Roasting ears" were another favorite, simply prepared by taking tender, young corn and roasting it in the husk in the coals of the fire or woodstove.

I have asked many Washington Parish old-timers and cooks to explain the difference between cornpone and cornbread. I have found that the answers and cooking techniques are as different as the people I interview. Generally speaking, cornpone is made from a simple mixture of cornmeal, bacon grease or other fat, and water or milk. Originally, it was something that could be made with few ingredients in the field, and fried on a black iron skillet, pot top, or even on a hoe blade, as "hoecakes." Thicker dough was shaped by hand, and thinner batter was poured onto a hot griddle or skillet. Cornbread, as you might expect, was often lighter, and might include eggs, milk, or buttermilk, and even baking soda and baking powder for leavening. Cornbread was also generally baked in a cast-iron Dutch oven in the coals of a fire, or in a cast-iron skillet in a woodburning oven. One thing nearly all Washington Parish cooks I've talked to seem to agree on is that sugar and wheat flour do not belong in cornbread or cornpone.

TRADITIONAL SIMPLE CORNPONE

Makes 4 servings.

— ◇ —

Many people ask me if there is a difference in flavor between white and yellow cornmeal. The truth is that, taste-wise, I really can't tell one from the other. However, I would argue there is a big difference in flavor when you use freshly ground cornmeal, whether coarse or fine. Many Washington Parish old-timers I've talked to have strong opinions about which grind is best. For cornpone, I like my cornmeal finely ground. My friend Homer Graves liked his very finely ground, but he recalled working at a mill growing up and serving one man, Mr. Buck Jarrell, who always wanted his as coarsely ground as chicken feed, like "biddie chops," as he would describe it. "When I eat mine, I want something to chew on," he'd say. When Jarrell was ninety years old he still had all of his teeth; according to Mr. Homer, "Mr. Buck didn't have no sto' bought teeth."

1 cup fresh, finely ground cornmeal

1 teaspoon baking powder

½ teaspoon salt

½ cup milk, buttermilk, or water

3 tablespoons bacon drippings, divided

Cane syrup for serving

1. In a large bowl, stir together the cornmeal, baking powder, and salt. Add the milk and 2 tablespoons of bacon drippings. Stir well and let sit 15 minutes.

2. Divide batter into four parts and hand press each piece into a small round pone about the size of the palm of your hand.

3. Place a large cast-iron skillet over medium heat and add the remaining tablespoon of bacon drippings. When the skillet is hot, cook pones until golden brown, about 3 minutes per side, flipping once. Serve warm with cane syrup.

LIGHTER CORNPONE

Makes 6 servings.

— ◇ —

These pones are light and fluffy, more like corncakes. They can be sweetened by serving with syrup, preferably cane syrup, with honey, or with fresh fruit (my favorite way to eat them). They can also be a savory side that is good with smothered pork chops. To make savory cornpone, add finely chopped green onion to this batter recipe.

2 cups fresh, finely ground cornmeal

1 teaspoon baking powder

½ teaspoon salt

1 large egg

1 cup milk

2 tablespoons bacon drippings, divided

2 tablespoons melted butter, divided

Cane syrup for serving

1. In a large bowl, stir together the cornmeal, baking powder, and salt. In a separate bowl, whisk the egg; then whisk in milk. To the liquid mixture, whisk in 1 tablespoon bacon drippings and 1 tablespoon melted butter.

2. Add liquid mixture to the dry cornmeal mixture and stir well. Let the batter sit 15 minutes, to let the milk soak into the cornmeal.

3. Place a 10-inch cast-iron skillet over medium heat and add the remaining tablespoons of bacon drippings and melted butter. When the skillet is hot, spoon about ¼ cup of batter into the skillet to form small pancake-sized pones. Cook until golden brown on each side, flipping once, about 3 minutes per side. Serve warm with cane syrup.

RUTHIE MAE GRAVES'S CORNBREAD

Makes 8 servings. Courtesy of Homer Graves.

— ◇ —

I have heard many debates about how to make cornbread and the best way to eat it. People I interview argue over how coarse or fine the cornmeal should be, or whether to use buttermilk or "rich" milk (fresh whole milk). Some are purists, never adding anything to their basic recipe, or they may enjoy the occasional "cracklin'" cornbread, made with crumbled fried pork skins mixed in. My late friend Homer Graves loved the way his wife Ruthie Mae made her cornbread, with just the perfect crispy crust, or "ruffles" as he called them. He would always say Ruthie Mae's cornbread could "make a man grit his teeth!" For supper, I like my cornbread crumbled in buttermilk.

> 2 cups white or yellow cornmeal (preferably
>   fresh and stone-ground)
> 1 teaspoon baking powder
> 1 teaspoon salt
> 1½ cups buttermilk
> 1 teaspoon baking soda
> 3 tablespoons melted unsalted butter, divided
> 3 tablespoons bacon drippings, divided
> 1 large egg, beaten

1. Put a 12-inch cast-iron skillet in the oven and preheat to 375°F. In a large bowl, combine the cornmeal, baking powder, and salt. In another bowl, mix the buttermilk and the baking soda. To the liquid, add 2 tablespoons melted butter and 2 tablespoons bacon drippings. Beat in the egg. Pour the wet ingredients into the dry ingredients and mix thoroughly.
2. When the skillet is hot, coat with the remaining tablespoons of bacon drippings and butter, keeping the skillet hot. Pour the batter immediately into the hot pan. Batter should sizzle as you pour.
3. Bake until golden brown, about 25 minutes. Serve warm from the pan, or remove and slice into triangles.

MASON JAR BUTTER

**How many it serves depends on how big your jar is
and how much cream you use.**

— ◇ —

My friend Bob Ann Breland, daughter of Beulah Mae Lang, remembers her childhood chore of making butter right before mealtime. She'd shake fresh cream in a mason jar until the butter separated from the buttermilk, which took a pretty good while. Bob Ann was never sure whether the purpose of this duty was to make the butter or to keep her busy while her mother made dinner, or both.

> 1 small mason jar with screw-top lid
> Enough heavy cream to fill the jar halfway
> Salt

1. Fill the jar halfway with cream and tighten the lid.
2. Shake the jar vigorously until lumps of butter appear, about 7–8 minutes. At this point there should be some liquid, the buttermilk.
3. Reach into the jar and press the lumps of butter into a ball. Strain everything through a sieve and reserve the buttermilk.
4. Rinse the ball of butter to soften it. Add salt to taste, working the salt evenly into the butter with your hands. Mold the butter into desired shape. (If there's enough buttermilk, make a treat by pouring it over crumbled cornbread in a bowl.)

Other traditional staples on family tables, which could keep starvation at bay during winter months, were dried beans, field peas, and butter beans. These legumes are not only filling, but they are highly nutritious, and are good sources of protein, fiber, and complex carbohydrates. They are also rich in vitamins and minerals, such as folates and potassium.

Beans and peas are not native to the American South. Field peas were originally brought to America by enslaved Africans. Butter beans, or limas, are from Mesoamerica and most likely came to the Gulf South region through trade conducted by native peoples in the area. Dried peas and beans served many purposes on Deep South farms. Not only did they feed farm families year-round, but they were seed stock for the next year's

planting. Field peas are also called "cow peas" because the peas were used as feed for livestock. Many farmers turned their cows out into fields that had been sown in peas as green fodder. Beans from the variety known as velvet beans, named for their velvety soft pods, were known to produce rich cow milk and cream, and it made good butter, or as Homer Graves called it, "that good ole cow salve butter."

Peas and beans don't need much attention to grow. They are extremely hardy, and they help rejuvenate the soil. I remember Mr. Seldon Lang told me, "You always want to plant those peas on the sorriest ground you've got. They can't stand prosperity." Peas and butter beans require little water and no fertilizer. "If you leave them alone," he would tell me, "they'll make little poor lookin' vines, but they'll be loaded with peas. If you put too much water and fertilizer, they'll make all vines and leaves and no peas."

Field peas were often planted in cornfields. The roots of these plants have nitrogen nodules that improve soil depleted by nitrogen-robbing corn crops. Toward the end of the growing season, peas were planted in the fall and early spring to produce what the old-timers called "green manure." They would plant the field peas, and when they were a foot or two tall, they would plow them under. The decomposing plants would build organic material in the soil, and that was fertilizer for the next crop. Many of today's modern crops require high doses of pesticides, fertilizers, and irrigation. In times past, however, farmers had to combine and rotate crops to keep soil healthy, and to attract bugs that could help keep harmful pests and parasites at bay. In winter months, farmers had to store beans and peas in ways designed to protect them from pests. Many old-timers talked about storing dried beans and other seeds in dried gourds, fabric bags, or paper sacks and coating the seeds with ground sassafras or wood ash to keep out weevils.

Field pea, or cowpea, varieties include crowder peas, purple-hull peas, black-eyed peas, and cream peas. Cream peas, also known as "creamer" peas and lady peas, are a real favorite in the Deep South because of their delicate, mild, and creamy texture. Butter beans, or lima beans, are another regional delicacy, especially when tender and shelled fresh. Limas come in two types: bush varieties, such as Jackson Wonder, or climbing-pole varieties, such as Purple Eye, Willow Leaf, or "Snow on the Mountain." Snap beans, such as the Rattlesnake Pole Bean or Louisiana Purple Pod, are also locally popular. Most people know and like regular old black-eyed peas, the most familiar type of field pea, but I just can't bring myself to eat them. There are so many

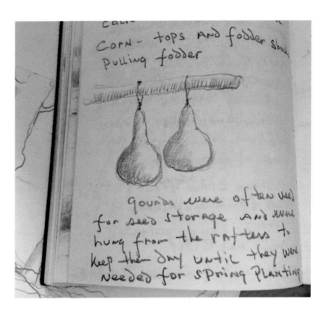

Many farmers grew certain types of gourds that were dried and used to hold seeds.

better-tasting varieties out there than the ones most folks buy in grocery stores.

Many bean seeds for growing are not widely available. But you can get a starter supply from organizations such as the Seed Savers Exchange, or online and through the mail from Baker Creek Seed Company and Southern Exposure Seed Exchange. For cooking, Camellia Brand Beans of New Orleans is an excellent source for dried peas, and you can find them online and in stores in most parts of the country. Of course, from time to time, I still find unique and treasured family-held varieties that are only shared among friends and family members from year to year. These are the ones I am most interested in safeguarding and using to increase the seed stock.

Before cooking, dried peas are washed, often soaked overnight, and then slowly simmered, most often with seasoned, smoked meat. The only universal tips I have

heard used in cooking peas and beans are to start the peas cooking in cold water and don't add salt to peas until they are done cooking, because it can make them tough. If you need more liquid in the pan, add boiling water to the simmering peas, not cold water, which will also make them tough.

FIELD PEAS OR LADY CREAMER PEAS
Makes eight 1-cup servings. Begin the day ahead.

—◇—

Many old-timers talk about eating field peas every day as children, or when times were lean, at every meal. As Mr. Homer Graves would say, "Give me a bowl of peas an' a pone of cornbread and I can plow my ole mule, 'Beck,' all day long." Field peas prepared in this traditional way are delicious, filling, and nutritious, especially when you have cornbread to dip in the pea potlikker. I like to eat mine served over rice. Sometimes I sprinkle them with salt pork or crumbled bacon.

1 pound dried peas, soaked overnight covered in
   2 inches water
10 thick slices smoked bacon, 1 smoked ham hock,
   or ¼ pound smoked, seasoned pork
1 medium yellow onion, finely chopped
2 cloves garlic, finely chopped
6 cups chicken broth or water
¼ cup thinly sliced scallions, with green tops
Salt and pepper

1. Rinse and sort the soaked peas. Set aside. In a heavy Dutch oven set over medium-high heat, cook meat until fat is rendered. Remove meat and pour off all but 2 tablespoons of fat. Cook onion in fat over moderate heat until softened, about 5 minutes. Add garlic and stir for 1 minute.

2. Stir in peas and broth. Bring to a boil, lower to a simmer, and cook, uncovered, until peas are tender, about 1 hour.

3. Stir in scallions, and salt and pepper to taste. Serve warm.

One of my favorite examples of an heirloom bean variety passed down through a family is the story of the "Snow on the Mountain" butter bean, a deep maroon, lima-shaped bean with white flecks that cluster along the inside curve and look like snow crests on a mountaintop. A fellow named Mr. Merritt from the Little Improve Community in Walthall County, Mississippi, gave it to me. He got it from a neighbor who was up in her late eighties. She got them from her grandmother, who, as a young bride, got them from her grandmother. That grandmother told her, "You keep this seed and you grow it. You take care of this bean throughout the rest of your life and it will take care of you. You will never go hungry. You can have those beans fresh starting in June, and they'll grow go all the way to Christmastime. If you are careful, you'll have dried beans to eat in the winter and you'll have plenty to replant come spring. They will keep you."

I think of the love those grandmothers were passing down with their seeds. And there's that young girl who kept that bean all those years, and this man passed them on to me. That's quite a gift. Anytime I get something that's part of a family's legacy, it's just a sacred privilege. Today we think of an heirloom as a photograph or a piece of furniture. That bean was a living heirloom that could mean a child's very survival, as well as a livelihood. I love what it illustrates: being self-sufficient and mindful of safeguarding your food, especially your seed stock. Relatively not too long ago, you had to preserve everything you'd need to live off of, and that had to be done year to year. Our forefathers were seed savers by necessity.

The "Snow on the Mountain" butter bean is one of my favorite varieties that I have collected over the years in Washington Parish.

1840 —
SNOW ON the
mountain Butterbean
from
Reverend Roy Blount,
Washington Parish
Louisiana 1991

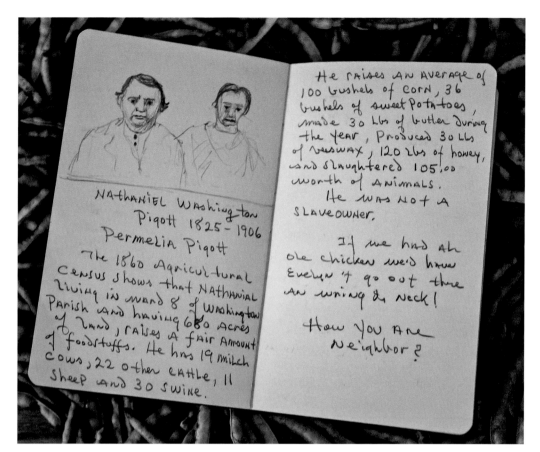

The 1860 Agricultural Census shows that Nathanial living in ward 8 of Washington Parish and having 680 acres of land, raises a fair amount of foodstuffs. He has 19 milch cows, 22 other cattle, 11 sheep and 30 swine.

Nathaniel Washington Pigott 1825-1906
Permelia Pigott

He raises an average of 100 bushels of corn, 36 bushels of sweet potatoes, made 30 lbs of butter during the year, produced 30 lbs of beeswax, 120 lbs of honey, and slaughtered 105.00 worth of animals.

He was not a slaveowner.

If we had an ole chicken we'd have Evelyn't go out there an wring its neck!

How you are Neighbor?

An October 5, 2008, journal entry recording the history of the Pigott family in the Washington Parish area.

Of course, sometimes family seeds were carefully guarded and not shared. One old family in the region, the Pigott family, grew a pea that was well-known for its superior flavor and creamy texture. The Pigotts settled in Washington Parish in 1837. The pea they grew was known as the Pigott Family Pea, and even though it probably had another name at one time, that was long forgotten. The old man of this family was proud of these peas and didn't want anyone else growing them. During one of my many Louisiana visits in the early 1990s, I was in the Circle-T Farm Supply Store in Franklinton, and a man named Clinton Miley heard me talking about old seeds, asking if anyone knew of any. He was right behind me at the checkout line and said he had "inherited" this old Pigott Pea from his wife's family fifty years earlier. He told me, "You follow me home, and I'll give you a start on that." So, I did and he gave me a sample of that Pigott field pea. That was pure luck.

Now, I can take that Pigott Family Pea or something like it and grow that out and have a bushel bag full of seed. I can put that in the freezer, and it's preserved and not technically extinct. But it's "functionally extinct" for anyone else, because they don't know I have it, and it's not available to them. To prevent practical extinction of

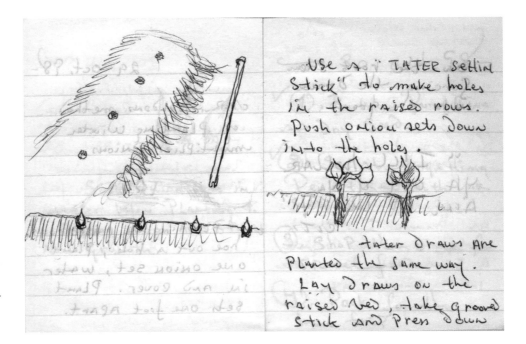

An October 29, 1998, sketch and journal entry with Seldon Lang's tips on growing winter multiplier onions.

the Pigott Family Pea, I have grown some on my land in Tennessee and registered it through Seed Savers Exchange where it is available to interested growers. I have also brought it back to Louisiana and handed it out to anyone who wants to grow it. That is what seed saving is all about.

Winter farm families also depended on things like multiplier onions, sweet potatoes, and pumpkins, which were stored in lofts and cribs. Pumpkins were typically put away in shuck cribs, where they were insulated from the cold by corn husks, and would keep until the following spring. A local farm favorite was the old-time tan field pumpkin, or the Upper Ground Sweet Potato Squash, which has a rich sweet flavor and makes delicious pies. Pumpkins were also commonly used as feed for cows and pigs, and many old-timers said they considered pumpkins as a good de-worming agent for their livestock.

Sweet potatoes served as another staple for many farm families in Louisiana and southern Mississippi. This highly nutritious, bright-orange root vegetable could be roasted in the hot coals of an open fire and served with homemade butter, sliced and fried, or mashed into potato pones. Many families kept baked sweet potatoes on hand as snacks between meals. During winter months, sweet potatoes were often stored in lofts and outbuildings or in trenches farmers dug out in the fields, where they were covered with pine straw and dirt to help protect them from freezing. They were also "banked up" in potato hills, buried with straw, and then covered with dirt and topped off with pine bark or a flat piece of tin that kept rain off. Some of the most popular varieties of sweet potatoes on Washington Parish farms were the Puerto Rican, the Jewel, the Dolly Varden, and the Dooley Yam.

" Them ole splinter torches we burn't ah many ah one of em when I wus comin' up, many's th night I got my lessons up uy th light ur ah splinter torch.

mamma laid them sweet taters on th hearth up agin th coals, an come supper time they wus ready fer that good ole cow salve butter.

February 16, 2015, sketches and journal entry from conversations with Arlie O'Bryant, who always talked about how his mother cooked on the open hearth, especially sweet potatoes.

ROSE VISE'S SWEET POTATO PIE

Makes 8 servings. Courtesy of Calvin Vise.

———— ◇ ————

You won't find much debate among Washington Parish cooks about whether sweet potato or pumpkin pies taste better. Most of the ladies I have worked with say sweet potatoes are the hands-down favorite.

> 2 cups peeled, cooked, and mashed sweet potatoes
>   (best if roasted)
> 3 large eggs
> ¾ cup sweet milk (whole, full-fat milk)
> ½ cup cane syrup
> 3 tablespoons butter
> 1 teaspoon ground cinnamon
> ¾ teaspoon salt
> ½ teaspoon grated nutmeg
> ½ teaspoon orange or satsuma zest
> 1 (9-inch) unbaked pie shell

Preheat oven to 375°F. Combine all filling ingredients in a bowl. Pour into pie shell and bake until firm, about 35–40 minutes. Cool completely before slicing.

Winter meals also relied heavily on canned vegetables and fruit from the summer growing season. Many old-timers remembered glass jars of all shapes and sizes that were kept indoors in pantries and kitchens where they wouldn't freeze during colder months. Mr. Arlie remembered having plenty to choose from: "During the winter we lived on what we put up during the summer and fall months. Mama canned peas, butter beans, tomatoes, okra, corn, and soup mixes. She'd have a couple hundred of them blue half-gallon canning jars filled with vegetables."

Canned okra, peas, corn, tomatoes, and green beans were often used in soups or cooked with stewed or smothered meat. Many vegetables—such as cucumbers, okra, green beans, and cabbage—were pickled to pre-serve crisp summer freshness. Canned tomatoes and tomato juice were used in a variety of dishes. I always tell people I would prefer a fresh canned tomato over a hot-house winter tomato any day.

BEULAH MAE LANG'S TOMATO GRAVY

Makes approximately 2 cups. Courtesy of Bob Ann Breland, daughter of Beulah Mae Lang.

— ◇ —

This is another one of my favorite recipes from Maw Lang's kitchen. I think it tastes best made from canned fresh tomatoes. It is delicious served over fried pork chops, rice, or a hot homemade biscuit split in half.

4 tablespoons bacon or pork chop drippings

3 tablespoons all-purpose flour

1 small onion, minced

1 (15-ounce) can of home-canned or stewed tomatoes, with juice

1½ to 2½ cups cold water

Salt and pepper to taste

1. Make a roux! In a heavy skillet, heat the bacon drippings over medium-low heat. Add flour and cook, stirring constantly, until medium to dark brown. Do not let the flour burn. (If you want, you can add the salt and pepper to the flour before you brown it.)

2. Add onion and stir constantly until onion turns translucent. Stir in the tomatoes and their juice, and stir around constantly until the tomato mixture is very thick.

3. Add 1½ cups of the water. Cook, stirring occasionally, until reduced and thickened, about 25 to 35 minutes. Add salt and pepper, and more water as needed. Serve hot over meat, rice, or biscuits.

There were also a good number of fruit preserves, such as wild crab apple jelly, Mayhaw jelly, blueberry, dewberry and blackberry jellies, and fig preserves. But most important were the gallons of cane syrup that would see the family through until the following fall.

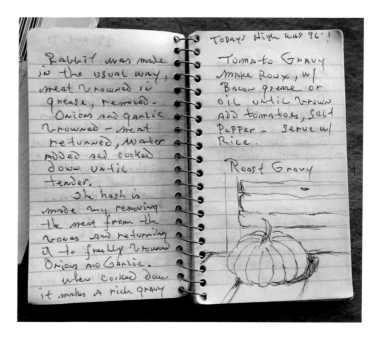

An October 18, 1993, journal entry with recipes I recorded at the Washington Parish Fair—rabbit hash and tomato gravy.

Many small farmers grew a patch of sugarcane, which was usually harvested in late November or December, when the sugar content in the stalks is highest. Cane was cut at ground level, collected, and crushed to extract the juice. Today this is done mechanically. Generations ago, syrup was extracted through a manually fed cane grinder powered by mules tethered to walk in a circle around the mechanism.

To make the syrup, juice from the crushed cane was slowly cooked in an evaporator for hours, then filtered to remove impurities. The cane syrup was stored in large metal tins and was used year-round. Even today, sugarcane is a major crop in Louisiana, comprising about 15 percent of sugar produced in the United States. Most modern sugarcane varieties have been genetically modified to increase the sugar content and make mechanical harvesting easier. A century ago, the most popular sugarcane for making syrup in the Gulf South was ribbon cane, a variety that grows between six and twelve feet tall, and that has a darker color and richer, less sweet flavor than today's sugarcane. The older cane varieties were softer, yielded more cane juice when crushed, and were often cut and chewed raw as a snack. Many old-timers told me that syrup was served at every meal. I have seen some Washington Parish residents cover everything on the plate with syrup.

SYRUP CAKE

Makes 12 servings. Recipe adapted from my journal notes.

— • ◇ • —

I first tried this cake at the Washington Parish Fair. It is a sweet treat that smells as delicious as it tastes, especially topped with a little powdered sugar and served with a cup of strong, hot coffee.

1 cup sugar

3 large eggs

1 cup cane syrup (I like Steen's brand made in
      Abbeville, Louisiana)

1 cup vegetable oil

2¾ cups all-purpose flour

2 teaspoons baking soda

1 teaspoon allspice

1 teaspoon cinnamon

1 cup hot water

1. Preheat oven to 350°F. Grease and flour a 9×13-inch baking pan and set aside. In a large bowl, beat together sugar and eggs. Add syrup and oil and mix well.

2. Mix in flour, baking soda, and spices. Stir in hot water. Pour into prepared baking pan and bake until a knife stuck in the center comes out clean, about 45–50 minutes. Serve warm or cold.

Wintertime also meant hog-killing time. Farmers raised hogs for meat, but their primary purpose was providing fat for the coming year, for cooking and other household needs. A hog butchering was one of the most important events of the year, and it involved all family members old enough to help. Once it turned cold enough to keep fresh meat air-chilled, usually late November or early December, the family would rise before the sun to stoke fires under large caldrons filled with water. After the pigs were slaughtered and bled, they would be scalded in boiling water to loosen the hair and bristles that could be scraped away from the skin. The cleaned hogs would be butchered and the meat cured. The fat was rendered to make lard. One old-timer told me, "only the squeal got away," meaning every part of the pig was put to use. Before the days of electricity, there was no way to keep uncured meat for an extended period of time, so freshly slaughtered cuts had to be used as quickly as possible. A special treat the night of a butchering was dining on the fresh tenderloins.

The curing process began as soon as butchering was completed. Hams, slab bacon, and every other cut of meat to be preserved were placed in poplar wood troughs and covered with salt to cure them before taking the cuts to the smokehouse. Mr. Seldon Lang recalled how his family tied hams and bacon to smokehouse rafters using bear grass, which resembled a smaller version of the yucca plant. They would run the sharp point of the bear grass through uncured meat and tie the slabs

" You could cook that on Ah lightard knot an hit'd be good!

You'd tie them SAUSAGES off and hang em on poles. Then you'd tie them midlins off with Bear with grass an then smoke em fer several days

ALL roads'll lead t Franklinton tomorrow, first day of the fair.

Hit's just ah ole dirt floor in that smokehouse. They'd burn hickory that-us would halb dry an half wet so hit'd smoke an not flair up.

A 2011 journal sketch detailing methods of smoking meat.

An October 24, 1998, journal entry:

The log smokehouse is filled with sausage hanging from rafter poles. Along with side meat and midlands tied with bear grass, smoldering hickory coals fill the room with hazy smoke, the old process of curing meat. Not far from the smokehouse two large kettles with oak fires under them are being tended by a group of men rendering lard. One kettle is finished and the lard is ready to be dipped into large tins where the lard will be stored for winter use. The second kettle is filled with square chunks of fat which will be cooked for cracklin's in the rendered lard. The kettles are stirred with large wooden paddles.

An October 2001 journal entry described the process for curing meat:

Cured Meat—We used pepper and red pepper mixed with salt and packed meat in the salt box for seven days and smoked it for three days. Packed hams in corn shucks in wooden barrels. We left the sausage hanging in the smokehouse. Momma used to fry sausage and pack it in a crock. A layer of sausage was put in the crock and then grease was poured over it—this was repeated in layers—we kept sausage all year that way.

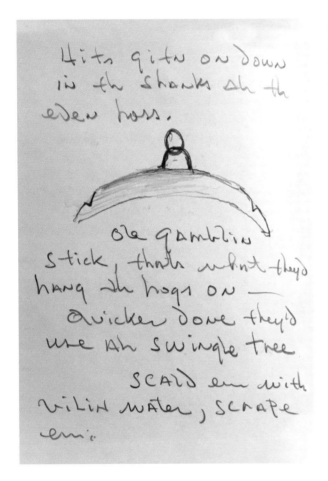

Hits gitn on down
in th shanks ah th
eden hoss.

Ole gamblin
stick, thats what theyd
hang ah hogs on —
Quicker done they'd
use ah swingle tree
scald em with
bilin water, scrape
em.

Sketch of a gambrel or "gamblin' stick," as my friend called it, used to hang meat in a smokehouse.

to the rafter poles in the smokehouse. Small fires were built on the smokehouse floor. Green hickory or pecan wood was added to the coals, and the smoke would fill the wooden building and give the meat that wonderful woodsy flavor. Much of the offal and rubber-tough cuts of meat were ground into sausage and mixed with salt, rubbed sage, and cayenne pepper flakes. The sausage

patties were fried and put into canning jars or ceramic crocks and covered with lard to preserve them all winter long. Families had to be careful to ration meat and lard, which had to last until the following year, when the process was repeated.

Homer Graves told me that when he was a child, he would sneak out to the smokehouse where the meat was stored and use his pocket knife to cut a few slivers of ham off the backsides, so that nobody would notice any was missing. "It didn't take mom long to find out, though. She asked me one morning, 'who's been cutting on my hams?' Whenever she did find out, she sure laid one on me."

The fat from the hog was cut into pieces and put into large cast-iron cauldrons that were placed over fires. The pieces were stirred with a wooden paddle until the lard was rendered. This process also produces a treat known as "cracklin'." Cracklin' is deep-fried fat with a little bit of skin attached. These fried pork skins are drained, cooled, seasoned, and enjoyed warm from the cauldron or stored for later snacking. Lard, the resulting liquid fat, was strained and allowed to cool, and was stored for use throughout the year. Long before cooking oils were available, farm families depended on lard to provide necessary dietary fats. Lard was used for frying, making biscuits and cornbread, and in numerous other household applications, including making salves and soaps.

Much of the winter's farm work centered on preparing fields and farm tools for spring planting. Winter fields would be "laid by," planted with late-fall cover crops such as winter rye, wheat, or clover. Cover crops rejuvenated the soil, crowded out weeds, and added new decomposing, carbon- and bacteria-rich organic material into the earth. As fields rested in the coldest months, preparations to use them again were already underway.

Mr. Homer Graves told me the end of one season meant the start of another. "After we picked the last of the cotton in December, we'd set in with our winter work getting the fields ready for plantin'. My Daddy'd have me grubbing out fence rows and choppin' sprouts out of the pasture fields. We used brar scythes t' mow down sprouts an' briars so we wouldn't get hung up when we was turnin' the mules out at the end of the row we was plowing."

Winter was also a time to repair and sharpen farm tools so they would be ready to use in the coming months. Mr. Homer remembered old Mr. Boss Spears, who was a well-known entrepreneur and blacksmith in Washington Parish. "Spears would be busy all winter making repairs on plows, beat'n out plow points, plowshares and all sech like that. We had to have all our farm equipment ready for spring plowing and planting. We'd replace plow stocks and hoe handles that was broke, sharpen hoes and plow points and have it already to go, come spring plowin' 'n plantin' time." Farmers most often made repairs themselves, and they used materials they had on hand or found in the woods.

I recall one special winter survey of Mr. Seldon Lang's property, during a visit in 1993. I was showing Mr. Seldon an old cotton-scraping hoe I had bought from Mr. J. D. Wise, now deceased, who had an antique stall at the Washington Parish Fair every October. This particular hoe head was one of the old types made with good metal that could be filed periodically to maintain a sharp edge for shaving off weeds and young sprouts.

Mr. Seldon explained that, unlike old tools, newer types of tools are made from thicker, heavier metal that cannot be filed and kept sharp. After filing and buffing the head of my "new" hoe down, Mr. Lang took me out in search of a handle. Walking through the woods, he recalled the time when his tool handles were hand-made from wood cut from nearby woodlands. Local folks would use several types of wood. Some of the most popular were white swamp bay saplings, sassafras, red sumac, and prickly ash. Those woods are light in weight, yet durable and less prone to cracking. The best of all, he told me, is the white swamp bay, and that is what we set out to find that day. Not just any white bay wood would do. It had to be just the right width, straight, and without flaws. Finally, after a good hike and much searching, we found the perfect sapling branch. We cut it and it carried it back to his workshop, stripped it of its bark with a drawknife, and fitted it on the hoe. The end result was a work of art—a beautiful antique hoe fitted on a newly made handle. This hoe remains a treasured relic from that day, and is a token of true friendship with Mr. Seldon.

Old-timers told me about how they would clear cut land for farming. What timber was not used for building was burned in large fires, a practice they called "burning bushes."

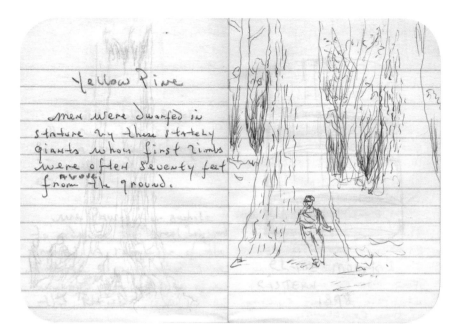

This is one of several journal entries from 1993 where I recorded Seldon Lang's memories of the yellow pine forests that once covered the Washington Parish landscape. He recalled hunting trips with his brother to an area of virgin forest along the Pearl River with great cypress trees along the bayou. Yellow heart pines towered overhead, forming a canopy so thick that sunlight was blocked out.

Another critical cold-weather task focused on what the old-timers referred to as "makin' new ground." This meant clearing wooded areas by removing trees and stumps, often using mules or oxen to drag them from the soil, and by burning or chopping out roots and underbrush. Mr. Arlie O'Bryant talked about how, as a boy in the 1930s, he could earn three dollars for clearing an entire field. It was backbreaking work but welcome, since money was difficult to come by, if there was any to be earned at all. Most often, tasks were performed as trades for sharecropping opportunities, tools, or seed. Mr. Arlie told me, "I bunched up logs for ole man Will McCaffrey and burnt 'em off to make new grounds. Mr. Will told me, he said, 'Son, you clean up th' new grounds, I'll let you make a crop on it.' And bless'd goodness, I made corn and cotton on them new grounds th' followin' year. On them winter days we'd bunch up brush and such and get it piled up 'fore we stuck far [fire] to it."

In the early years of Washington Parish's history, the yellow pine forests were so dense that homesteaders could not possibly use all of the cut timber for building purposes. Toward the turn of the nineteenth century, homesteaders began clearcutting the forests to make farmland. This type of work drew the community together for what was known as a "log-rolling." Several families would gather to fell large stands of pine trees and roll them to an area where they could be cut up or carried to a sawmill. The men and boys would work in teams with mules to drag the logs out of the woods. Smaller children would gather, pile, and sort brush to be burned or reused for kindling and the garden. While the men worked clearing the trees, the women would gather to mend and sew quilts for people in the community, usually newlyweds, new mothers, and older people. The families that worked together would bring enough food to feed the entire group dinner. It was a full-day project.

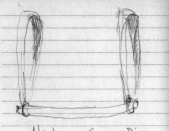

Hickory Cross Piece,
rings and sewn doubled
Sack cloth apparatus
for lifting logs.

General Rester
" man out plowin', " then
Nats' after em, swarmin'
around his rear end.
       They's ah log rollin'
6 men + a log, George
said let's see what that
little feller Kim lift
can tote. Hit's so
heavy the sunk in hard
ground.

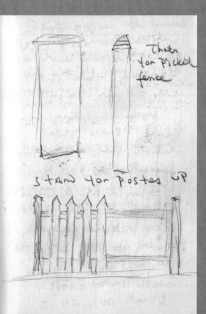

they made them
picket fences out a
Pine. You cut th
log as long as you
wanted it then you'd
rive th boards out

1st cut

split th log
in 1/4s

Thats
Yor Picket
fence

Stand Yor postes up

Log-rollings were major community events to help clear land for farming, cut lumber for sale and for building. Many of the old-timers I interviewed described the tools that work teams used to roll and drag heavy logs, like this diagram from a 1995 entry, of cross pieces with sack cloth looped to make harnesses.

Logs were also split to make everything, from boards and fencing to wooden shingles. Even branches were gathered and used to construct supports for the garden. Wood chips and scraps were used to make tar, in tar pits that were dug into the ground. An October 17, 1995, journal entry described the process: "They would dig the pit out in a cone shape that slipped down to the center. Then they would dig a trench out of the pit and put a barrel at the end to catch the tar. The pit would be 10 feet high around the rim and the lighter knots would be stacked in the pit and set on fire where it would burn for days."

Lightard Section Line Post on a Washington Parish back road in the early 1900's.

There are two Post Oaks on either side of the road.

Small dirt roads through the back country were common during the early years of the Parish's history

A 1993 sketch and journal entry about the immense lightard stump Mr. Seldon and I discovered on an afternoon walk. Lightard is a colloquial term for resin-rich pine heartwood prized for its burning qualities.

After the log-rolling, the timber would be used or sold by the land owner, or it would be bunched up to be burned. Pinewood was used for building and repairing houses, barns, outbuildings, roofs, and fences. Homer Graves recalled his family's board-and-batten house and its wood-plank roof. "I remember that we had snow in those winter months when I was coming up. We had a split board roof on our log house. They was 3-foot boards that was split-off yellow pine. If we had a snow storm come in and the wind was blowing hard, snow would find its way through any cracks or spaces in that house. I slept under several quilts on them cold winter nights and I have woke many of ah mornin' t' find snow on my top quilt."

Split logs were often used to build fences and fence posts that kept free-roaming livestock out of cultivated fields and marked property boundaries. Split wood was also used for firewood, and brush was saved for kindling. Mr. Calvin Vise would split logs in half and quarter them. He would then take a drawknife, which had a handle on each end and a blade in the middle, and he would rive splits off, a half-inch thick. He would bend and tie the pliable wood into supports for climbing bean plants. Even brush branches were collected to fashion bean supports for growing plants in the early spring. Other wood scraps were slowly burned in a special tin kiln to make tar. Nothing was wasted.

After the fields were cleared of trees and the stumps were removed, there was still more work to be done to get the fields ready to plant. Roots and other brush had to be dug up to prepare the ground for spring plowing. Experienced farmers knew that "new ground" was extremely fertile soil, and often, as soon as the field was ready, winter greens were planted on the spot.

At Christmastime, many of my friends recalled how, as children, they hunted for the perfect tree in the pine forests. "We'd go out somewhere's along the edge uv' ah ole fence row into the woods and cut us a yeller pine that was the old longleaf pine. We'd hunt us one that wuz 'bout six foot tall," recalled Mr. Seldon. "We'd put that tree up in the front room and hang paper icicles on it, then we'd put a few ah them colored glass balls on th' tree t' finish it out. Most folks back then didn' get ah lot in th' way uh Christmas presents. People just didn't have th' money t' buy presents with. On Christmas morning,

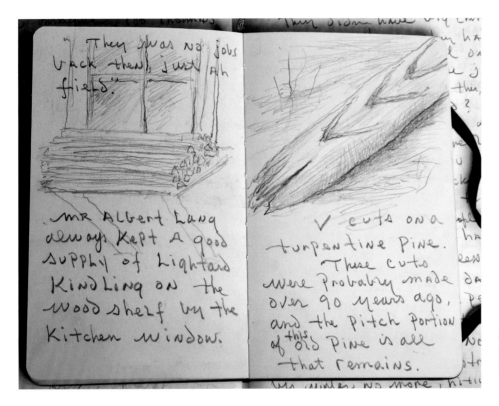

Handwritten notes in sketch (left page): "They was no jobs back then, just ah fields." "Mr Albert Lang always kept a good supply of lightard kindling on the wood shelf by the kitchen window."

(right page): "V cuts on a turpentine pine. These cuts were probably made over 90 years ago, and the pitch portion of this old pine is all that remains."

A February 17, 2005, sketch showing Vs cut in trees for harvesting turpentine.

laying by my britches, would be ah few apples an' oranges. Then we'd have raisins in boxes. We generally got sum candy t' go along with the rest of it. Our Christmas dinner was always chicken an' dumplins with chicken dressing."

In the late 1800s, Washington Parish's stands of virgin pine drew national timber companies to the piney woods region, where more than a million acres of longleaf pine were logged out by the 1930s. The industry drew a diverse collective of workers from all over the Southeast, some of whom scraped together enough of their meager wages to purchase cheap plots of clear-cut land. This may have accounted for a higher number of small farms, particularly ones owned by African Americans and European immigrants, than in many other

parts of Louisiana during the same time. This land ownership, combined with proximity to rail lines built by the timber companies in the 1930s, also gave rise to a significant truck-farming industry that supplied fresh produce to restaurants and communities in New Orleans to the south, and Memphis and Chicago to the north.

The timber stands were also treasured for another resource. Seldon Lang was a young boy when he first saw trees being tapped for their resinous sap. Turpentine camps, in areas sometimes referred to as "groves," would spring up wherever the work was being done. Mr. Seldon recalled that the difficult, transient work would draw large teams of migrant laborers. Entire families would often follow the work team and live in the camps. Many old-timers told me their families made money selling food and goods to migrant turpentine workers. Mr. Red Fussell recalled his father selling chickens, butter, eggs, syrup, and sausage to families in the work camps.

Turpentine harvesting required strong men to move through stretches of virgin forest and cut grooves into the trees with a special tool called a turpentine hack. The tree-cutting tool, which weighed about twenty-five pounds, had a grooved blade in the shape of a V on one end of a two-foot long handle. On the other end was a weighted metal ball that you would hit with a hammer to carve a V into the tree trunk. As the crew worked its way through the grove, each man followed a line or section. A number of V-shaped cuts were made on both sides of the trunk of every tree being tapped. At the

An October 22, 1996, journal entry about hunting.

base of each tapered cut, a small trough-shaped piece of metal was inserted. Below this trough, two pegs were driven into the tree, onto which a box-shaped metal container was placed to catch the dripping sap.

As the containers filled with sap, teams of mules pulled flatbed wagons carrying empty barrels that would make their way through the groves. Men with long-handled dippers dipped the sap into the barrels. After the barrels were filled, the wagons returned to the turpentine still, where raw sap was distilled into turpentine.

My own first-hand experience with turpentine lore came on a late fall afternoon when Mr. Seldon and I were walking through the woods in a swampy area near Witch Creek. This area outside of Franklinton is not far from his farm, and on that particular afternoon we were out in search of lightered wood. When we reached the right place, Mr. Seldon pointed out a section of an old pine trunk whose bark had long since vanished, leaving the wood's worn surface smooth from years of weathering. Carved into the wood were a number of V-shaped cuts that had been pounded in by workers over ninety years earlier.

In the late nineteenth and early twentieth centuries, turpentine had many uses. A refined version of it known as "rectified spirits of turpentine" was used medicinally. Mrs. Letha Toney told me that when she was a child and had a cold, her mother's remedy was a tablespoon of castor oil with two drops of rectified turpentine spirits. Mr. Seldon recalled the use of red pine oil, which was a derivative of the distilling process. Red pine oil had a beautiful rich color, which was a deep reddish amber when held up to the light. It was used to treat cuts, sores, and sprains. These home remedies were often the only medicine available. Of course, while some people still rub turpentine oil on sore joints to ease pain, we now know it should never be taken by mouth. "We had to do our own doctorin' back then," Mr. Arlie O'Bryant would recall. Even old-time fiddlers used to get their bow resin from yellow pine trees. They took sap from a cut in the tree and molded it into a cake. After the resin dried and hardened, it was used on the fiddler's bow.

The pine forests and cypress swamps in Washington Parish supported large populations of game and wildlife that were a significant food source for farm families during winter months. These animals also provided meat and pelts that could be sold. Unlike today, in the 1800s and early 1900s, large game such as deer was scarce. But squirrel, rabbit, opossum, and raccoons were plentiful. "They was times when that was about all you had to eat," Mr. Seldon Lang told me. He was an avid hunter, as was his father, and many family meals centered on wild game. They did a lot of market hunting and trapping along the creeks, and out in the swamps, too. "We'd trap whatever we could that would bring us a price," he remembered.

Because the family rarely had access to wheat flour to make brown gravy, Mr. Seldon's mother prepared most game with what the family called "streaked" gravy, pronounced "streek-ed." Mr. Seldon's wife, Beulah Mae, learned the technique from her mother-in-law, who was part Choctaw. This gravy was the family's favorite dish, and she prepared it often. She began by browning whatever game she was cooking in bacon grease or shortening in a cast-iron Dutch oven. She'd remove the meat and add onions and garlic, and sauté them until soft. Then she would add water to the drippings and add the meat back to the gravy and simmer until tender. She made squirrel and rabbit this way. Maw Lang usually prepared venison roast for Sunday dinner or holiday meals. Mr. Seldon's brothers, who lived nearby, could often smell this dish cooking as the heavenly scent wafted down the road, and would drop in for dinner.

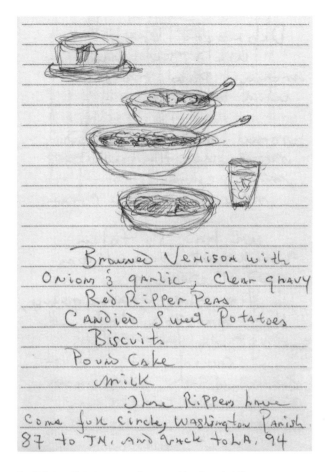

An October 18, 1994, menu for a meal I enjoyed at the Lang family table.

It is often said that a dog is a man's best friend, but for farm families, dogs were a necessity. Many farmers owned small dogs they called "feists," or hounds that were prized for their abilities to track or tree wildlife. Mr. Seldon Lang often talked about his dogs. "You had coon dogs, mink dogs, squirrel dogs, rabbit dogs. You had huntin' dogs of all kinds. I had a red bone hound and a curl tail squirrel feist." These dogs were also a source of pride and bragging rights.

While traveling to Louisiana, I would often bring my dog, Nicholas. Nicholas served as a willing companion.

Mr. Arlie O'Bryant claimed he once had a dog that could point out a bird with its right paw and a rabbit with its left. One of his hunting buddies doubted the dog's abilities. During one hunt, Mr. Arlie and his friend followed the pup out of the swamp and into a cotton field. "That feist threw up her right paw and we jumped a covey of quail," Arlie told me. "Then we passed into a cornfield, and that feist stopped and threw up her left paw. And, sure enough, there was a rabbit. After seein' that, that doubtin' feller asked me, 'How much you want for that dog?'" Mr. Arlie told his friend, "She ain't for sale."

In many rural areas of America, there is an entire culture centered on sharing hunting stories. Washington Parish is no different. These stories and other family tales still provide hours of entertainment around bonfires, fireplaces, and dinner tables, especially during winter months when days are short. In earlier generations, certain members of the community, like Mr. Seldon Lang, were celebrated for their storytelling craft, for their ability to turn a phrase, to build suspense, and to make a captive audience laugh. Before electricity and the advent of radio and television, these conversations allowed people to connect and share family histories, community news, and to just enjoy each other's company. During my visits to Washington Parish, I still sit for hours and listen to my old friends' stories and philosophical reflections with all of their genuine goodness and simple humor. I continue to record everything I can.

One such story, told to me by Mr. Seldon, dated back to a winter day in his childhood when he was working alongside his father. The two were setting lightered fence posts at the far end of their cotton field, using post-hole diggers to cut through the sandy gray soil. Suddenly they were interrupted by sounds coming from the nearby woods. The woodland floor was covered with dry oak and hickory leaves which accentuated the wrestling

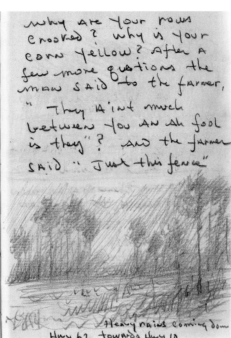

well Uncle John, hits Such rememberances of ole Hun.

A fellow was Plow'n out corn on a hillside sideling one day and a man came along and sked, "why are those rows crooked? and the man Plow'ng said they were straight when I Plowed them, the sun warped them.

One day a city man stopped at the fence where a farmer was Plowing his corn crop and he began to ask questions.

why are your rows crooked? why is your corn yellow? After a few more questions the man said to the farmer, "They aint much between you an ah fool is they"? and the farmer said " Just this fence"

Hwy 62 towards Hwy 10 Heavy rains coming dow-

This 2006 journal entry captured the skepticism some rural farmers felt about city dwellers.

crackling sound of some type of small animal bounding through the leaves. Pausing from their work, Seldon and his father turned toward the woods just in time to see a large swamp jackrabbit leap from the underbrush into the cotton field and begin a zigzagging course toward the road some several hundred yards distant.

Scarcely had the rabbit disappeared into the cotton field when they heard the sound of a much larger creature barreling through the woods, chasing behind. Limbs were being brushed aside, twigs broken, and dried leaves kicked as something or someone drew closer to the spot where they were working. Suddenly the underbrush parted and their neighbor, Mr. Zeke Thomas, appeared in midair, having just cleared a large pine log hurdle with shotgun raised high in his right hand and the left arm pointing straight out for balance. Mr. Zeke cleared the log with all the grace of a track-star hurdler, landing

just a few feet from the spot where father and son were standing. "The rabbit went that-a-way!" Mr. Seldon's father yelled as he pointed across the cotton fields next to the road, followed by a question for the hunter, "How you gonna feed them ten children on that one rabbit?" As he passed by on the run Mr. Zeke replied, "It's not the rabbit I'm after, it's th' gravy!"

My journals contain hundreds of stories like these. While I could not possibly include all of them here, I do have some favorites featured throughout this book, such as the story of the traveler from New Orleans who stopped to talk to a farmer plowing his corn crop. He began asking question after question: "Why are your rows crooked? Why is your corn yellow?" After a few more questions, the man asked the farmer, "There ain't much between you and a fool is there?" The farmer replied to the man, "Just this fence."

A 1994 sketch of telling stories around the fireplace.

Storytelling was essential around the hearth, especially when it turned cold. "Come winter time," Mr. Arlie O'Bryant would tell me, "we'd sit around the fireplace and listen t' them ole folks tell tales about them ole days an' ole ways. I did hear my grandpa tell that our people come from that island across the great water, he said they was from County Cork [in Ireland]."

One of my favorite things to do with old-timers in Washington Parish was sit around a fire and listen to the stories of their childhood experiences. "I did my studying by coal-oil lamp or by light from the fireplace. I come up poor, John. I knowed I was hungry 'cause when I ate breakfast I was usually eating warmed over peas with potlikker poured over the cornbread. It was a hard life, but it was a good life. It was a simpler life, simpler way of life." Mr. Arlie would say, "People weren't scattered then. You didn't have time to be scattered," he said. "Over yonder you had a tater patch that needed plowing out, across the road you had a pea patch that needed laying by, over in the next field you had corn to plow out. There was something all the time. And that was your way of life."

Winter holidays in Washington Parish served as a time to rest a little. It was a homecoming for families, and an occasion for special meals, especially as younger family members have moved away. On Christmas Day, many friends recalled looking forward to a venison roast, wild turkey, or chicken and dumplings. During wintertime holidays, large family gatherings in this part of Louisiana often feature what they call a pig roast, which involves slow-roasting a suckling pig over a smoldering fire.

I always find the turning of the calendar year makes me reflect on the natural seasonal cycles of life and the inevitable passage of time. For many of these old farm communities, it was a time to celebrate the fruits of their labor with family and friends before the planting cycle started again. Today, we still gather with family and friends to reflect on the end of one year and celebrate the promise of another. Observing holiday traditions inspires us to look backward, often with nostalgia.

All too fast, the ways and lives of our ancestors have faded before us like the embers of a smoldering fire. We have often traded their old ways for today's "improved" modern conveniences, which have obvious benefits. But as I sit and visit with folks from an earlier generation, I can't help but wonder what we lost in that trade. These days, many of us struggle to recognize the circadian and seasonal rhythms our bodies crave. But as a farmer, just as I know spring rains bring the promise of new life and new growth out of dormant ground, I find comfort and a unique sense of peace in observing traditions of the past around a table, surrounded by dear friends and family, sharing our lives in the simplest way.

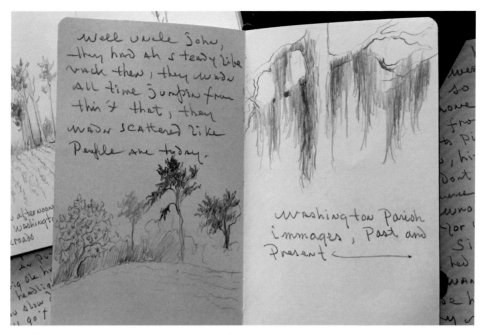

October 2007 sketches and a journal entry recording Arlie O'Bryant's thoughts about how life has changed.

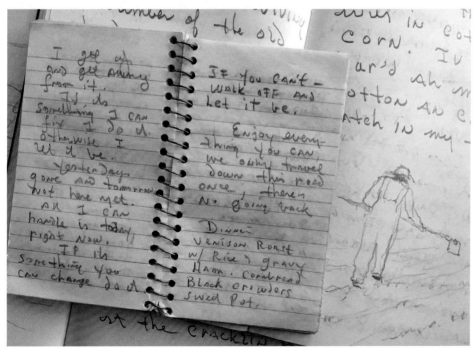

A February 12, 1990, journal entry from conversations with Seldon Lang, who always reminded me to live each day to the fullest.

SPRING

You can always tell when spring is coming. There is a certain smell that marks the change to this season. It is an earthy, microbial aroma of damp, fresh-turned soil and new green sprouts. The sun begins to heat blankets of dew into a morning mist that lifts from a waking landscape. The sounds change too. As the ground begins to warm, the quiet stillness of winter is replaced with the occasional chirp of a bird, a croaking frog, or the hum of insect wings.

Up until the mid-1950s, when the ground began to shake off winter's chill, young farm boys like my old Washington Parish friends and their fathers before them would harness their mules, hitch their plows, and head out to turn and smooth the soil one last time before planting. From an early age, they did the work of men. As Mr. Arlie told me, "I done my share a' plowin'. I was pushin' a middle splitter when I was thirteen years old. That mule had t' turn 'round at th' end a' the row with that plow, 'cause I couldn't lift it." Winter cover crops "laid by" over the colder months had nourished the fields and were plowed under, building the rich organic material that continually nourished the ground. Farmers dragged their plowed fields with a harrow, much like a large rake, that would break up the ground and roots into finer soil. Then they'd begin a delicate series of passes over the land with individualized tools that created the ideal hills and furrows needed to plant seeds and catch rain water.

My friends often talk about the tools they had at their disposal for working the land. Today, these early farm implements are rare finds. I've tried to sketch as many as I can, when I am fortunate enough to come across them. To prepare fields for planting, farmers used mule-drawn plows with interchangeable points. These plow points had descriptive and utilitarian names such as the bull-tongue, duck-bill, middle splitter, buzzard sweep, heel or "hill" sweep. There were also the spring tooth, drag and disc harrows, which many of my friends refer to as "hars." One of Mr. Homer's most prized possessions was what he called a "pope-side har." After a little digging, I learned that this plow's namesake, William Winston Pope, was a US congressman from Mississippi who lived not far across the Louisiana state line from Homer Graves. In 1886, Pope designed and patented a highly useful side harrow with teeth that were "adjustable up and down and sidewise." Mr. Homer always said this was one of the most important tools he owned. There were

Enduring Seeds

In these fields the
seeds of our ancestors
have been grown for as
long as any living family
member can remember.
With the return of
spring every year, the red and
black field peas seeds are
sewn in the warm sandy
gray loam to grow once again
as a part of our "living
history and heritage".

Mr. Seldon house
seen from the old homesite
In the days when this
land was farmed, one long
field stretched from where
I am standing all the way
to the highway

An undated sketch and journal entry about
Seldon Lang's work on his family farm.

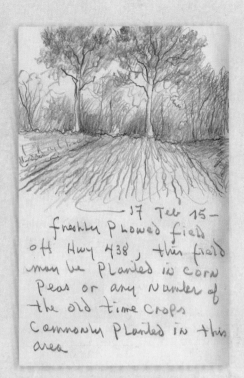

17 Feb 15 —
freshly plowed field
off Hwy 438, this field
may be planted in corn
peas or any number of
the old time crops
commonly planted in this
area

I often sketch Washington Parish farm fields ready for planting.

An October 2000 journal entry:

With the beginning of each new season I am reminded of those who've gone before us, it is their seeds that we sow in the warm soil of the early spring. This year the heritage garden at Blackberry Farm contains four heirloom varieties given to me by Mr. Seldon Lang, my long-time farming friend from Washington Parish, Louisiana. Mr. Seldon died a year ago this past May, but his heritage crops live on. I am reminded of this through the growing season as I watch these links with our past grow from seedlings to seed-bearing plants. Cowhorn okra and red and black field peas remind me of dinners in Maw Lang's kitchen and the Pearl River County Mississippi pre–Civil War peanut handed down through a family that descended from slaves.

wagons and "ground slides," which were basically makeshift wagons without wheels, and they looked sort of like a sled and were dragged over the ground by a horse or mule.

Of course, the most valuable tool on the farm was a mule. Up until the middle of the twentieth century, some farmers in the region had oxen and a few had horses. But if you didn't have a mule, you were in trouble. "I never did plow a horse much," Mr. Homer told me. "I plowed mules. The mules could stand th' heat better than a horse could." Mules did all manner of work on

the farm, and were ably handled by boys as well as men. "We took better care of our mules than we did ourselves. We had to, our livelihood depended on 'em. I plowed them mules 'til they turned white. When I come outta th' field of a eden [in the evening], them mules had enough salt on 'em ta cure a side uh bacon."

Trained to work, mules were essential partners, but they were also known to have personalities of their own. Mr. Arlie talked about working mules that knew when it was quitting time. At the end of a long day, the animals would instinctively head off for the barn, bouncing a

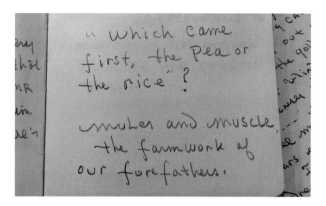

A 2014 journal entry recording one of the thousands of humorous exchanges and sage bits of wisdom and insight imparted to me by the older farmers I have met, as well as an observation about the hardworking nature of our ancestors.

plow through a field and busting through fences, even if the work wasn't finished. Others dutifully worked alongside hardworking farm boys, such as Mr. Homer, whose favorite mule was named "Ole Beck." There was an art and a rhythm to the work the two did together during the long, hot, tedious days.

I have spent many long days in the fields with my Washington Parish friends, helping them plant their gardens. And I have often imagined what it would have been like to start the farming year alongside these men in their youth, when there was more work to be done than hours in the day. I've been told that my English peas would have been planted in the ground in early January, alongside the last of my greens, cabbage plants, onions, and a few other cold-tolerant varieties. My Irish potatoes, or "arish taters" as my Louisiana friends call them, would have had to be in the ground by Valentine's Day. Once the threat of frost passed, they planted crisp lettuces. When tender spring produce started to come up, it often immediately made its way to farm tables.

KILT SALAD

Makes 6 servings.

———◇———

This is an early spring salad I grew up eating in the Smoky Mountains of Tennessee, and I was surprised to find that my friends in Washington Parish served a similar version. I've always called it "Kilt" salad. But it is really a spring lettuce salad that is wilted by a warm bacon-dripping vinaigrette.

1 large or 2 small heads of spring lettuce (I like
    Bibb or buttercrunch)
3 green onions, green and white parts
4 slices thick-sliced bacon
2 tablespoons bacon grease (reserved from
    cooking bacon)
2 tablespoons apple cider vinegar
2 tablespoons water
1 tablespoon brown sugar
Cracked black pepper to taste

1. Break up the lettuce and thinly slice the green onions. Toss them together in a bowl and set aside.

2. In a small, heavy skillet set over low heat, cook the bacon until fat is rendered. Remove bacon from the pan and drain on paper towels. Leave 2 tablespoons of drippings in the skillet. When bacon is cool enough to handle, crumble it and toss into salad mix.

3. Make a dressing by heating reserved bacon grease over medium heat. Carefully add vinegar, water, brown sugar, and black pepper. Stir until bubbly.

4. Pour warm dressing over salad mixture and toss gently. Cover bowl with aluminum foil or a pot lid for a couple of minutes to allow the lettuce to wilt. Serve hot.

A May 27, 1989, journal entry describing how often I thought about my friends in Louisiana when I was back home in Tennessee:

Many times at home while working on my farm I would think of my friend and teacher Paw Lang and wonder what he was working on at that moment. Then I would think of how it would be when I was down here and we were working side-by-side in the fields again, learning from his vast experience and knowledge—the living history of a bygone era. Few remain now to pass on knowledge from those times past and perhaps just as few to receive it now and pass it on.

ENGLISH PEAS AND NEW POTATOES
Makes 6 servings.

—◇—

One of my favorite early spring dishes was tender English peas and new potatoes in a buttery, white cream sauce. There is nothing more delicious then tender "new potatoes," so named because they are tender, immature potatoes with skin so thin that it literally peels off as you wash them.

1 pound small new potatoes

1 pound fresh or frozen small green peas

2 tablespoons butter

2 tablespoons all-purpose flour

1½ cups "rich" milk (fresh whole milk)

Salt and pepper to taste

Thinly sliced green onions for garnish

1. Cut unpeeled potatoes into quarters and put them in a medium saucepan. Cover with water, bring to a boil, and cook 12 minutes.
2. Add peas and cook until potatoes are tender, 5–7 more minutes. Do not overcook. Drain and set aside.
3. In a large saucepan set over medium heat, melt the butter. Add flour and stir constantly until you have a white roux, about 5 minutes. Whisk in milk and allow to thicken.
4. Add cooked potatoes and peas and simmer another 5–6 minutes. Add salt and pepper. Toss with green onions and serve.

Sweet potatoes were another staple that went into the ground in spring, when these farmers planted what they called "tater draws," or potato slips. Mr. Seldon probably taught me the most about the cycle of growing sweet potatoes, techniques he learned from his father.

He told me to start in late January or February with what he called "bedded up" sweet potatoes. This means he would dig a shallow trench, line it with fresh manure, and top that with a layer of pine straw. Then he scattered whole sweet potatoes over the straw and piled them up with another layer of straw and dirt. The manure generated heat and helped kick-start the potatoes sprouting.

By mid-March or the beginning of April, Mr. Seldon said those bedded-up potato sprouts would be ready to plant. He'd remove the covering from the beds, break the sprouts, or "draws," off the potatoes, roots and all. Then he covered the potatoes again to grow more draws. Mr. Homer recalled taking large baskets full of tater draws out into the field to plant.

On one of my earliest visits to Washington Parish, during the late 1980s, Mr. Seldon fashioned for me what I consider another prized possession. One morning while planning our day's work planting sweet potatoes, I lamented the arduous task of planting or "setting out" sweet potato draws. I had always performed the back-bending work by digging a small hole for each draw, filling it with water, and covering the draw with dirt. Mr. Seldon walked over to a wall in his toolshed and took down what he called his "tater-settin' stick," which was a simple but practical implement that eliminated tedious sweet potato planting drudgery.

The stick was a smooth piece of wood about four feet long, one-and-a-half inches wide, and three-quarters of an inch thick. One end was rounded off to form a smooth handle. The other end tapered down to a flattened point, about three-quarters of an inch wide. In the center of the tapered point, a rounded notch had been cut and smoothed down so it wouldn't cut or bruise the tater draws when planting. Mr. Seldon and I set out into the woods in search of a stick to make another one. After a long search, he found a perfect, straight stick. We

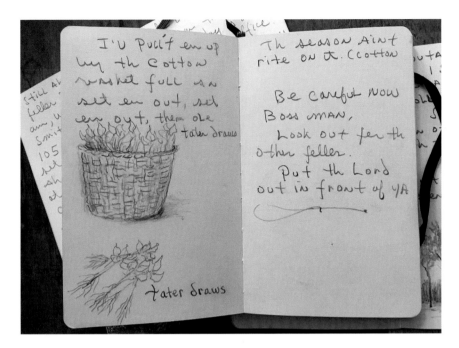

Large baskets of tater draws (sweet potato slips) were taken out to the fields to plant.

returned to his woodshed, where he hand-carved a "tater stick" for me. I still have it displayed in my garden shed, when I am not using it.

I can still hear Mr. Seldon's instructions: "We'd drop them tater draws about a foot apart on them bedded-up rows. Then we'd push the rooted end of the tater draw in the ground with tater-settin' sticks." Even today, when I plant sweet potatoes, I prepare my field with furrows and six- to eight-inch-tall raised rows. I lay my tater draws along the top of the row, just like he taught me. Then I use the grooved point of the tater stick he made for me to gently push the rooted end of the draw down into the soil, until the sprout stands upright on top of the hill. If all goes as planned, the growing sweet potatoes will be ready to harvest by late summer or early fall.

Early spring was also the time to prepare the corn-fields. "When it came time t' get corn out," Mr. Arlie told me, "Daddy always said if you didn't have yer corn out

by the end ah March, you was late." The seed would be "dropped," by hand, two grains to a hill, a foot apart in the row. You'd follow with a "sweep" that would throw a thin layer of dirt over the seeds. When the corn started coming up a few days later and the plants were big enough to work around, they'd use harrows and hoes to scrape weeds out from beneath the sprouting corn. Next they would "run the middles out," meaning they'd drag harrows between the rows and scrape out any remaining weeds with hoes.

Cotton, another important cash crop in Washington Parish, also went into the ground by the middle to the end of March. The planting process was similar to plant-ing corn. Mr. Seldon told me his father, Mr. Albert Lang, would plant cotton using the signs of the zodiac, ideally planting when the signs were in "the twins," claiming that it made double bolls of cotton on a plant. Cotton seed was drilled into the soil with a wheelbarrow-like

A 2000 journal entry:

Back years ago, Daddy would drop watermelon seed in the cotton rows after it was laid by. Come cotton-picking time, we'd find a watermelon every so often in the field when we were picking cotton. It was a real treat! In my daddy's time we would plant a terrace in watermelons. He used manure from the mule stable. Generally, we aimed to have our first melon by the Fourth of July. That was from the time that they were seeded 'til harvest.

cotton planter that was pulled by a mule along a row. When cotton was four to six inches high, farm boys used hoes and elbow grease to "chop out" the cotton, or thin it to about a foot or more between plants in a row.

By April, after the corn and cotton crops were in the ground, these farmers prepared to plant other summer crops, including beans, butter beans, tomatoes, summer squash, and melons. Stakes and supports made from brush collected over the winter were set out for tomatoes and beans to climb. Once those food crops were in the ground, it wouldn't be long before those first field peas and beans were ready to pick.

Generally speaking, produce gardens for the table were tended by the women on the farm. And when families visited one another, one of the first stops was a tour of the garden. Mr. Seldon recalled that his mother plowed her garden with an ox. In this long-ago age, one admiring visitor remarked to his father, Mr. Albert Lang, "That's the prettiest work I ever saw done that wasn't done by a man or a mule."

One of the first treats of spring was strawberries. These sweet, delicate berries grew wild in the acidic soil of the piney woods region and were gathered by indigenous peoples and early settlers. In the late 1800s, immigrants from Italy, Sicily, and Hungary, along with small-patch farmers, first started cultivating strawberries on areas clear-cut in timber operations on the north shore of Lake Pontchartrain, north of New Orleans. By the early part of the twentieth century, the Florida Parishes were a major supplier of strawberries to commercial American markets along the rail lines. Well-known varieties like the Klondike, Tangi, and Daybreak were small, misshapen, and deep bright red, with a higher sugar content than most modern commercial strawberries. Under threat of extinction today, a select few growers located around Tangipahoa and Livingston parishes

Henry Harrison, retired LSU agricultural extension agent.

are working to preserve these Louisiana Heritage strawberry varieties.

My good friend Henry Harrison recalled growing and harvesting the juicy berries on his aunt and uncle's farm as a young child in the 1950s: "We mulched the berry plants throughout the growing season with pine needles that we gathered in the woods. Then we would get up before dawn to harvest the berries before school and be back out picking after school. We packed crate after crate and loaded them on cold-storage railcars. During harvest season, hundreds of these railcars snaked through the tiny town of Independence, loaded with strawberries headed for New Orleans restaurants and grocers up north. They were so delicious, but you'd be so sick of them by the end of the season." I first met Henry Harrison in 1990, during one of my Louisiana visits. Today, he is the semiretired Washington Parish agricultural extension agent for Louisiana State University and is active in the region's garden clubs and agricultural life. For someone who is semiretired, he keeps awfully busy. He is still known as the region's resident expert and go-to man for any problems growers have with their crops. Over the years, he has been an invaluable source of knowledge for me on local, heirloom seeds and unusual varieties of fruits and vegetables, not to mention his farming know-how.

STRAWBERRY PRESERVES

Makes approximately five 8-ounce jars.

—◇—

Louisiana's piney woods region was once famous for its sweet varieties of strawberries. When the crop came in in the spring, the berries were quickly picked and shipped out to grocers all along the Illinois Central Railroad. What wasn't shipped had to be preserved quickly, to capture peak freshness. Berry preserves could be used year-round with biscuits and cornbread, or as a sweetener in jam cakes.

5 pounds fresh strawberries (preferably the smaller varieties grown on local farms)

2½ pounds raw cane sugar (more, if strawberries are the commercially available varieties)

5 tablespoons freshly squeezed lemon juice

1. Sterilize five 8-ounce glass jars and lids. Put berries, sugar, and lemon juice in a large, heavy-bottomed saucepan and stir constantly over low heat until the sugar is dissolved and berries become soft and juicy.

2. Increase the heat to medium. Stirring occasionally, boil until the mixture reaches 210°F, about 30–40 minutes.

3. Use a sterilized ladle to spoon the hot mixture into prepared jars. Process jars according to manufacturer's instructions.

Spring also brought the first of the year's fresh beans and peas, which were prepared in a variety of ways. Beans and peas were staple food crops for farm families in Washington Parish up until the middle of the twentieth century. The food traditions that came out of this dependency are still very much alive. As you would imagine, field pea plants contain peas in all stages of growth throughout the growing season. One regional first dish of the late spring and early summer was the standard "peas and snaps," which makes use of the new, tender, and delicate bean "snaps," named for the sound they make when you break them, combined with the first tender green shelled peas from more mature pods. This dish is usually simmered with some kind of smoked meat for seasoning. Another favorite is green beans or snap beans simmered with new potatoes in chicken broth, and seasoned with onions and smoked meat. Of course, with so many heirloom varieties grown in Washington Parish, the beans and peas themselves provide a wide variety of flavors.

Along with finding unique varieties of beans, peas, and other crops, I also come across sayings, figures of speech, rhymes, and other things related to them. I always write down any details that are of particular interest, whether it's information about the family that passed something along to me or maybe even one of their favorite recipes. For instance, I love this poem about watermelons: "Eat the meat. Pickle the rind. Save the seeds for plantin' time." One of the best things I ever recorded was a song that an old-timer taught me called "Just a Bowl of Butter Beans," sung to the melody of the beloved hymn, "Just a Closer Walk With Thee."

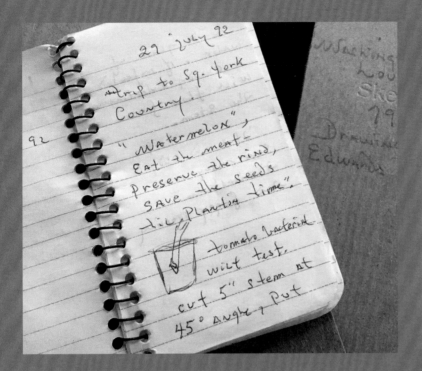

A July 29, 1992, journal entry featuring a poem I heard about watermelons and a sketch to test tomato plants for damaging bacteria.

A 2006 sketch and description of butter beans from the Washington Parish area.

A 2014 journal entry recording the lyrics of a song Homer Graves would sing for me called, "Just a Bowl of Butter Beans" to the tune of the old favorite hymn, "Just a Closer Walk with Thee." Mr. Homer's lyrics were a little different.

BEULAH MAE LANG'S BUTTER BEANS
Makes eight 1-cup servings. Courtesy of Bob Ann Breland.

Dried Reverend Taylor Colored Butter Beans.

I was first introduced to this method of cooking butter beans by Mr. Seldon Lang's wife, Beulah Mae Lang. "Maw Lang," as I came to call her, was a wonderful cook. She learned much of what she knew from her mother-in-law, Narcissus Simmons Lang, who was part Choctaw. Maw Lang would start with tender, shelled, immature speckled butter beans, because Mr. Lang didn't like them after they had matured and were in fact speckled. She'd cook them with real butter and full-fat milk. She always served them with biscuits, never cornbread.

1 pound fresh or frozen young, tender butter beans
or baby limas
1 quart water
3 tablespoons butter
2 tablespoons of "rich" milk (fresh whole milk)
¼ teaspoon freshly cracked black pepper, or to taste
Salt to taste

1. Rinse beans and place in a pot with water. Heat to a boil; then reduce temperature to a simmer. Cook, uncovered, until beans begin to get creamy, about 45 minutes.
2. Stir in the butter and milk, and bring to a simmer. Cook, uncovered, until beans are soft and creamy, about 15 more minutes. Add pepper and salt. Let stand 10 minutes before serving.

In this part of the country, biscuit making is a true craft. In all my visits to Washington Parish, and of all of the biscuits I've enjoyed there, I cannot recall any two batches of biscuits being the same. The older cooks I have known each have their own special touch or version and approach to biscuit making. No one uses recipes. Instead, one generation teaches the next about approximate ratios of measurements and consistency of biscuit dough by feel.

Mrs. Melba Hershey told me that if she was making biscuits to "pull," or drag, through cane syrup and butter, she would work the dough a little longer so that the biscuits would hold up and not crumble. But then she would add, "don't work the dough too long or they'll come out packy." Presumably, she meant on the tough side.

BEULAH MAE LANG'S BISCUITS

Makes 8 biscuits. Courtesy of Bob Ann Breland.

———◇———

My good friend Bob Ann Breland is an accomplished columnist and food editor for the *Era Leader,* the local newspaper in Washington Parish. Her description of her mother's biscuit making sums up the process for me: "She sifted flour into the dough bowl (no measuring except by eye) and then made a well in the center. In the well, she placed a "hand full" of shortening (like Crisco) then added a 'glass full' of milk as needed as she mixed the biscuits." Of course, Maw often used a sour-milk dough starter that she kept in the refrigerator. She made biscuits so often, it never went bad. Here is my best attempt at measurements for her recipe, with the option of using buttermilk instead of her starter.

2 cups self-rising flour

⅓ cup, or a "handful" of lard or vegetable shortening, as Maw Lang would call for

1 cup milk or buttermilk

1. Preheat oven to 500°F. Sift flour into a large bowl and make a crater in the middle. Add the shortening into the center of the crater. Gradually add the milk, working the shortening and flour together by hand.

2. Pinch off biscuit-size pieces of dough and flatten to about ¾ inch thick. Put a slight thumbprint indention in the bottom of each uncooked biscuit.

3. On a greased cast-iron griddle or heavy baking sheet, place uncooked biscuits in a circle, touching, with one in the center. Immediately put biscuits into preheated oven. Bake until golden brown, about 12–15 minutes. Serve warm.

HOPPIN' JOHN

Makes ten 1-cup servings. Adapted from a recipe by Mrs. Ventress Young from the *Covered Wagon Cookbook,* published in 1959. Courtesy of Washington Parish Fair Historical Society.

———◇———

Hoppin' John is a well-known dish across the American South, most often prepared with field peas. Where there are peas, there is bound to be rice to go with them, and I certainly found that to be the case in Washington Parish. As with many recipes used by busy farm families, this old-time method of making Hoppin' John required little attention. This allowed the cook to tend to other things while the peas and rice simmered in a Dutch oven in the coals of the fireplace or oven.

4 slices bacon

1 medium onion, chopped

4 cups water

1 cup raw field peas

1 cup raw rice

Salt to taste

1. Preheat oven to 350°F. In a heavy Dutch oven that has a lid, cook bacon until soft. Remove the bacon from the grease and reserve drippings. Chop cooked bacon and set aside.

2. Sauté chopped onion in reserved bacon drippings until tender. Add water and peas. Heat to a boil, reduce heat, and simmer until peas are tender, about 40 minutes.

3. Strain peas and reserve pea potlikker. Measure potlikker and add water to make 2 cups liquid.

4. Return peas, bacon, and potlikker to the Dutch oven and stir in uncooked rice. Put the lid on the Dutch oven and place in preheated oven for 1 hour. Take out of oven and let sit a few minutes with the lid on. Serve warm.

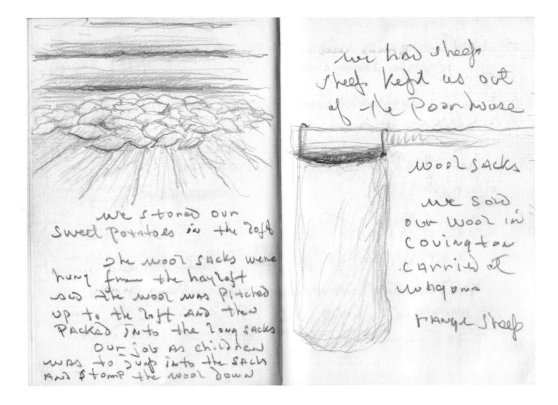

A spring 2001 journal entry detailing some of the ways my friend's families made money during the Great Depression.

Rural life in the early twentieth century wasn't easy. While farmers were busy planting crops that would feed and provide for their families, they were also always on the lookout for ways to earn a bit of extra money to purchase things they could not grow, such as flour, coffee, and white sugar, as well as farm tools and equipment that they couldn't make themselves. The local general store was one of the few places to buy these things with cash, through barter or by providing lines of credit. One of my favorite memories from all of my trips to Washington Parish was my first visit to the Varnado General Store in downtown Franklinton. The store had been opened in 1906 by timber dealer, bank director, and land speculator Daniel Sheridan, and it sold everything from farm supplies to household goods, clothing, and shoes. The Varnado family bought the store in 1920 and operated it until the mid-1980s. In 1973, I bought a cast-iron pot with a lid, a tin water dipper, and a white enamel dishpan. I still have all three items. Every time I look at them, I'm reminded of what it must have been like to purchase essentials from the local general store, when resources were scarce.

Aside from selling cotton, which was harvested in late summer and early fall, Washington Parish farmers often kept hogs, cattle, goats, and poultry. Some even raised sheep for wool. Most farmers raised an array of poultry for their eggs, meat, and feathers. Chickens were prized possessions, landing on the family table for

Sunday meals and when company came to visit. Fried chicken, roast chicken, chicken pies, and chicken and dumplings usually signaled a special occasion. Chickens were generally free to roam until they nested down in coops or around the farm at night. Once in a while, a neighbor's stray chicken might wander into another family's yard and seek shelter beneath their house. One old-timer recalled that, as a child, his brother would tie grains of corn on a string and dangle the corn through cracks in the floor boards. The dancing corn kernels would lure an unsuspecting chicken to a weak area in the floor where he could remove a loose board, snatch the chicken, wring its neck, and dispose of the evidence discretely and deliciously, without the neighbor ever knowing what happened.

Some birds did double duty. Mr. Arlie and Mr. Seldon both talked about keeping guinea hens. Guineas are loud, flocking birds that like to perch high in treetops. They look like polka-dotted vultures with smallish featherless heads that are almost shaped like a peacock's. They served many purposes on farms. A guinea's loud cry warns when unwanted predators are near. They eat harmful insects and are known to attack intruders, such as snakes and other egg eaters. Many old-timers also talked about how delicious guinea fowl tasted.

ROSE VISE'S STUFFED EGGS
Makes 24 stuffed eggs. Courtesy of Calvin Vise.

— ◇ —

3 slices bacon

12 hard-boiled eggs, cooled and shelled

2 tablespoons mayonnaise, preferably made fresh;
   recipe on page 130

2 tablespoons grated onion

1 tablespoon prepared mustard

2 teaspoons hot sauce

1½ teaspoons Worcestershire sauce

Salt and pepper to taste

1. Cook bacon. Drain and crumble.

2. Slice eggs in half down their length and put the yolks into a mixing bowl. Arrange the empty egg whites on a serving plate. To the yolks, add remaining ingredients and crumbled bacon, and beat with a mixer until smooth.

3. Use a spoon or pastry bag to fill the egg-white halves with yolk mixture. Serve immediately, or cover and refrigerate.

A bird many old-timers recalled raising on Gulf South farms was a now-rare breed known as the Cotton Patch Goose, which served as nature's lawn mower in a cotton patch. These small, tawny gray, orange-beaked birds would be turned out in the field to eat grassy weeds that could choke off cotton plants. Children would steer the geese through the cotton field with sticks to make sure they did their job. Unlike most geese, these didn't migrate, and they could handle southern heat. The Cotton Patch Goose also required little care or attention, and was prized for its meat and eggs. Fortunately, today select breeders are trying to revive the species across the southern United States.

Another old breed raised by Gulf South farmers was a variety of range livestock known as Pineywoods cattle. Mr. Homer described them for me: "I remember them ole range cows they had here. They had big old long horns on 'em and you'd see 'em out of the woods and swamps. They was a lot a range hogs out there too, and some folks had their cattle and hogs branded so they could identify 'em come roundup time."

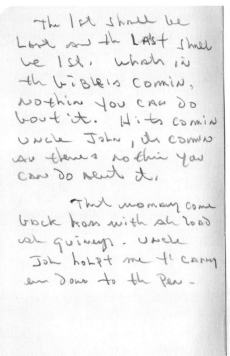

A 2006 sketch and journal entry describing Mr. Arlie's philosophy on life and his pens of guinea hens.

Pineywoods cattle descended from Spanish cattle and are known for their straight or slightly upwardly curved horns, similar to Texas longhorn cattle. They are extremely heat-tolerant and, through natural selection, they were relatively disease- and pest-resistant. In the old days, they were often left to forage in the area's forests and swamps. Many old-timers told me that when spring rains created high water in the swamps, they would have to rescue their livestock by boat. They would also drive their cattle ninety miles south to rail lines or boats that took them to markets in New Orleans and beyond. Those same railroads helped the area around Washington Parish become a leader in the state's dairy industry. Today, Pineywoods cattle are considered an endangered breed, and select ranchers in small pockets of the southern and central regions of the United States are working to protect them.

Traditionally, farmers kept some cows for milk and hogs penned for meat. Most often the livestock were let loose and allowed to feed in the swamps, woods, and laid-by fields, keeping farmers from having to use their dwindling supply of feed and milling corn. Hogs provided the largest source of meat and fat for farm families. "Now they was a lot of 'em that was plum wild," recalled Mr. Homer. "Some of them wild hogs you see out of the swamps today are descended from them hogs way back yonder." Today, many parts of the South have tremendous problems with feral hogs. Feral hogs are omnivorous, prolific, and have no natural predators. They are a real menace, and can be especially damaging

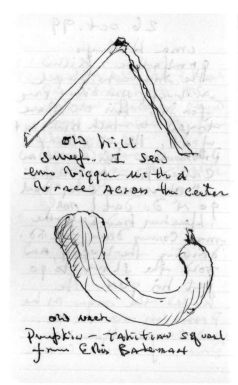

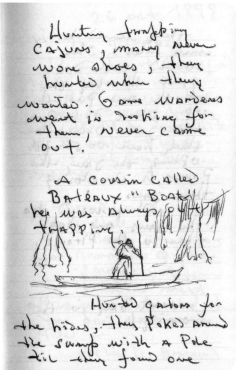

A 1999 journal entry detailing an old plow head, a rare squash variety I found, and local hunting and fishing culture.

to crops and food sources consumed by other wildlife, like deer. Many hunters across the state of Louisiana and parts of the Deep South still hunt for wild boar, which are butchered for their dark, rich meat, which makes flavorful smoked sausage.

During the spring planting season, hardworking farmers had little time for anything but work, but there was always time to go fishing. One of the most welcome signs of the season was budding trees, a sure clue that the fish would be biting, and that after a long day in the fields, it would be time to grab fishing poles. Old-timers say that certain types of tree buds can predict when certain fish will likely bite. Oak buds mean the bass are biting, and sprouting pecan leaves mean it's time to try your luck at catching bream and white perch, also known as crappie. Of course, catfish are plentiful almost year-round, but spring is one of the best times to catch them.

Many old friends talked about wrapping up a long workday by packing up blankets and fishing poles, lines, and traps to go camping along the banks of nearby rivers. Mr. Arlie told me stories of fishing trips where they'd build a campfire on the creek banks and stuff sweet potatoes in the coals while they went "mudcatting," as they would call catching catfish. They'd often place glass "minnow jars" in streams to catch bait. Sometimes at night, the farmers ran fishing lines or set out basket-weave traps, which they hoped would catch fish while they were back in the fields the next day.

PAN-FRIED FISH

Makes six ⅓-pound servings.

Adapted from my journal notes.

———◇———

When there was time to fish, panfish were always favorites at farm tables and creek-bank cookouts. Old-timers said they'd often stay up most of the night fishing and telling stories by campfire on river banks. There is nothing much tastier than fried fresh fish.

> 2 pounds fresh white perch or other small panfish (or filets of larger white, flaky fish such as catfish), thinly sliced, no more than ⅓-inch thick
> 1½ cups milk
> 2 large eggs
> 2 cups very finely ground cornmeal
> ½ cup self-rising flour
> 1 teaspoon cayenne pepper or
>   2 teaspoons Cajun-style seasoning
> Salt to taste
> Peanut oil for frying (2 inches deep in pan)
> Special equipment: a temperature gauge

1. Rinse fish filets in cold water and pat dry. Beat together milk and eggs. Place fish in egg mixture and refrigerate at least an hour.
2. In a large, shallow bowl, mix together cornmeal, flour, cayenne pepper, and salt.
3. Heat oil in a Dutch oven or deep fryer to 350°F. Shake egg mixture from the fish filets and dredge them in the cornmeal mixture. Carefully place in oil and fry until golden brown. Drain on paper towels and serve hot.

FRIED CORNMEAL DUMPLINGS

Makes 6 to 8 small dumplings.

Recipe from my journal notebooks.

—◇—

While some places call these "hush puppies," or "hot water cornbread," I have never heard them called that in Washington Parish. These are particularly tasty with fresh fried fish or fried pork chops.

    1 cup fresh, finely ground cornmeal

    2 tablespoons all-purpose flour

    1 tablespoon baking powder

    1 teaspoon sugar (optional)

    ½ teaspoon salt

    ½ cup boiling water

    1½ tablespoons finely chopped green onion (optional)

    Pinch cayenne pepper (optional)

    Bacon grease, peanut oil, or vegetable oil
        (¼ inch deep in your skillet)

1. In a large bowl, mix together cornmeal, flour, baking powder, sugar, and salt. Add boiling water and mix. Mix in green onion and cayenne pepper, if desired.

2. Add bacon grease to a cast-iron skillet and heat it over a medium-high fire. While grease heats, let the batter sit. It will thicken into a stiff dough.

3. Pat dough into small palm-sized pones and carefully drop them into the hot grease. Fry until golden brown on the bottom, about 3 minutes, and flip with a slotted spoon. Fry until second side is brown, 2 to 3 minutes more. Drain on paper towels or a paper plate. Serve warm.

At the beginning of the twentieth century, another favorite pastime in this farm community was spirited get-togethers known as "frolics." Family, friends, and neighbors would turn out for what was essentially an all-night dance party. Mr. Seldon often talked about these events. Come Saturday night, the host family would clear the furniture out of the main room of their home to make space for the dance floor and guests. Entire families would clean up and dress in their best, load their wagons and buggies, and ride several miles over rough, unpaved roadways for the night's frolic. Adolescent farm boys would court girls they only ever saw in church or school and would take the opportunity to socialize and to dance, which was also called "stomping the pea vine." Families brought enough food, sweet treats, and refreshments for everyone, which would be set out on long tables in the room.

A 1999 journal sketch of fiddle players entertaining audiences at the Washington Parish Fair.

BEULAH MAE LANG'S TEA CAKES

Makes 4 dozen cookies. Courtesy of Bob Ann Breland.

— ◆ —

These cookies were among my favorites when I visited the Langs' house, and they would have been a welcome treat at any frolic. They were usually served warm and fresh out of the oven with cold raw cow's milk. As with most of her specialties, Beulah Mae Lang did not use a recipe. This is her daughter Bob Ann Breland's best guess.

> 2 cups sugar
>
> ½ pound (2 sticks) butter, softened
>
> 3 large eggs
>
> ½ teaspoon vanilla extract
>
> ½ teaspoon freshly ground nutmeg (optional)
>
> 3 to 3½ cups sifted self-rising flour

1. Preheat oven to 350°F. In a large bowl, cream together sugar and softened butter. Stir in eggs, one at a time.
2. Add vanilla and nutmeg. Make a well in the center of the mixture and start adding enough flour to make a dough stiff enough to roll out.
3. Shape dough into approximately 1-inch balls and flatten each with a fork or the bottom of a glass. Place on a lightly greased cookie sheet and bake until just before they get brown, about 9 to 10 minutes.

Center stage at the Saturday night frolics was a group of fiddle players who performed the music for dancing. As Mr. Seldon described it to me, "One fiddler would play the tunes, and another musician would 'beat the straws,' the term for using small reed canes to tap out the rhythm of the song." The way this worked was that,

A 1994 sketch of moonshine and fiddlers "beating the straws."

while a fiddler played, the percussionist would take his canes, or "fiddlesticks," and drum on the neck of the fiddle strings between the fiddler's bow and his fingering hand. Another man could "call each set," or direct the dancers' moves by shouting steps and partner combinations to rouse them into a frenzy. Mr. Albert Lang was known as a great fiddle player in his day. It was said he could play the fiddle all night long and never play the same song twice.

Along with the food, no party was complete without jugs of moonshine, or "white mule" as it was known. The heavier the white mule flowed, the more feverish the dancing. "As the evening progressed," Mr. Seldon recalled, "the air of inebriation heightened and the fiddlers' strings loosened." The musicians played favorites from the time, such as "Turkey in the Straw," "Sally Gooden," "Billy in the Low Ground," "Black Mountain Rag," "Saturday Night Waltz," and many more.

When the sun came up, the fiddler would play "Home Sweet Home" to end the evening.

A frolic was a grand occasion for everyone. Children who came with their families busied themselves in the yard playing games such as hide-and-seek and rolling hoop rings. Sometimes they'd jump and slide on hay mounds in the barn loft, where they would tire out and eventually drift off to sleep. Back at the party, older generations would sit along the walls and watch the action or gather on the front porch, sharing news and storytelling. As late as the frolics often stretched into the night, there was still no excuse for not showing up spit-shined for church on Sunday morning.

Another rowdy community tradition involved performing what was referred to as a "chivalry," which is the local way of spelling and pronouncing the word "shivaree." When a young couple got married, friends and family would celebrate the wedding night by showing up outside the newlyweds' door and carrying out any number of pranks to drive them outside to salute the crowd. Mr. Lee Hill, a frequent visitor to the Washington Parish Fair, talked about his memories of chivalries: "I chivalried many a one of them back in the days. I mean we kept up some kinda racket. We'd go to the newlyweds' house beatin' on tin tubs, ringin' cowbells, hittin' wash tubs, and playin' cornstalk fiddles. We'd tie a string to the side of the house and stretch it tight to a tree and covered it with yeller [yellow] pine resin. When you'd flip the string, it made the awfulist racket you ever heard."

When the newlyweds finally emerged from their wedded bliss to plea for quiet, Mr. Seldon recalled that the groom was hoisted up on a fence rail and carried around the house several times. The bride would be placed in a wheelbarrow and pushed around the house, through the woods, and back again. When the newlyweds finally met each other back on the front porch, the rowdy crowd serenaded them with joyful songs and fiddle playing as they passed a jug of white mule to toast the marriage. Finally, as the merriment wound down, intoxicated revelers would head off for home, their melodies fading into the night, finally leaving the newlyweds to celebrate their honeymoon.

There were many ways the community bonded together to help newly married couples get off on the right foot. In her nineties at the time, Mrs. Eileen Crane told me, "When a young married couple was ready to set up, family members and neighbors would join together for a log-rolling. They'd clear the trees off the couple's plot of land, and the logs were hewn and notched to build them a log house. I remember three-foot yellow-heart pine boards were rived off [split] for the roofing. After they finished the house, they'd build that couple a barn and outbuildings."

So many of these old-timers talked about a time when people really depended on each other to make it. In the years following the Civil War, Mr. Seldon's grandfather lived about sixty miles north of the town of Franklinton, up around the town of Magnolia in Pike County, Mississippi. Following the war, people survived on little more than field peas. Even if there was livestock, salt was scarce, making curing meat difficult. It was a common practice to remove dirt from the smokehouse floor, put it in a cast-iron washpot, and add water. This mixture was boiled for a time and allowed to cool. Once the dirt had settled to the bottom of the pot, the fat rose to the surface and was skimmed off and used for baking bread or seasoning peas. What they were actually using was the fat and drippings that had fallen to the dirt floor when hams, bacon, and other cuts were being cured with hickory smoke.

Parched corn was another survival staple during those lean years. Corncobs were burned, and the white ash that remained was used for baking soda. For making bread, sometimes white oak acorns were used to make flour. Acorns were first soaked in the creek several days to remove their bitter acidic taste. Then they were dried or parched, and ground or pounded into a type of meal. Many people foraged in the woods for wild grapes, berries, and nuts. They did what they had to do to survive.

The same hardships struck during the Great Depression, when there were no jobs or extra money to be made. As in most of America, farmers in the region had to make the best of what they had, and they had to help each other grow what they could. Mr. Seldon's father often noted that there were a lot of bad cooks during the Depression, "because they weren't nothin' to cook."

One of the most memorable stories from this time was told to me by Mrs. Opal Breland, as she and I volunteered together at the Washington Parish Fair. She told me how desperate things were for her family during the Depression. There was no money, and it was all they could do just to put food on the table. Opal's daddy had died when she was little, and it was up to her mama to feed the large family, which included her grandparents and four siblings. "Mama was the farmer in the family, and everyone looked to her to make a crop to see them through," Miss Opal recalled:

It was an unusually early spring that year, so she had plowed the field, gone over it with a drag harrow, and laid the rows off in anticipation of getting an early start on the season. There were many mouths to feed and not a day to waste. All went well at first. The corn was dropped and covered. Following several days of spring rains and warm weather, it was starting to sprout.

Then one day it happened. Almost without warning, the weather changed and a heavy freeze came in from the North. In a single night, the crop that Mama had worked so hard to start was wiped out. A walk across the field revealed a most unnerving sight: the newly sprouted corn lay black on the plowed earth. Desperation at once began to set in. There was not enough seed corn to replant the lost crop, and neighbors had none to spare either. Mama thought out loud, "How am I going to feed my family? How will I make a crop?"

Without corn, there could be no meal for bread, one of the main staples that sustained the family, and there would be no feed for the stock either. Opal remembered that her mother began to pray: "She prayed and prayed for a miracle to take place, and she prayed on for days after that. She prayed for a miracle that would save

her family from starvation, but as the days past, the prospects for deliverance seemed more remote. On the morning of the second week, Mama walked out to the field to offer up one more prayer for salvation. When she reached the field, she could hardly believe her eyes. Here and there among the long rows, tender green shoots were beginning to appear. Mama fell to her knees and offered a prayer of thanksgiving for a true miracle."

In the days that followed, more green shoots appeared until at last all of the rows were evenly germinated and a good stand of corn was assured. Not only had the crop been spared, but the season proved to be the most successful ever. When harvesttime came in the fall, the family gathered and pulled more corn than they could use, and the surplus was given to two neighbors who were in need. Stories like this between generations, emphasizing the power of faith and perseverance, could bolster spirits in tough times and illuminate life's blessings.

—◆—

I have collected a number of varieties of seeds from food crops that have been historically grown throughout the South, mostly corn, beans, and peas. I have found many of them near my own home in the Tennessee foothills of the Smoky Mountains, and many are from southern Mississippi and Washington Parish, Louisiana. I have been told that I possess a remarkable ability to identify each of the seeds I've collected by sight and to recall, from memory, the smallest details about each one. I don't know how extraordinary that talent is, but I equate my knowledge about the seeds I collect to adopting a child. Each one is precious and unique, and I'm responsible for it. I won't be the one who allows any harm to come to it while it is in my care. As my friend Mike Lang, Seldon Lang's son, says, knowing details about seeds is just like knowing your own grandchildren: "You have to know each one's name and birthday."

One of the reasons I rely so heavily on memory for details about seeds is because when I started seed saving back in the early 1960s, there were no networks, computers, or databases to track the information. Outside of the universities and the US Department of Agriculture, which focused primarily on breeding more mass-producible, commercial varieties of seeds, there was not really anyone doing seed-saving work in any coordinated sense. In other words, at that time, if you were saving and documenting old seeds down in Louisiana, I would have no idea you existed, and you'd have no idea that I existed in Knoxville, Tennessee. Fast forward to today, when we have a great networking system. Many times when I'm talking to guests at Blackberry Farm or in Washington Parish, if I mention an old tomato or some old variety, someone will pull a little phone out of their back pocket and in seconds they've got all sorts of pictures and information about it. What I would have given for that gadget in 1959 when I was looking for so many elusive varieties.

I am no wizard of technology. I rarely use my own cell phone. I don't email, and I wouldn't know how to Google something to save my life. But I do know the benefit of using the Internet in seed saving and connecting like-minded people. The Internet is a tremendous resource for learning how to save and safeguard heirloom seeds, how to register, catalog, and share them, and how to purchase and grow old varieties properly so they remain pure to type and don't cross strains.

When we say a plant is an "heirloom" variety, that technically means it has been around for at least fifty years, and it is open-pollinated or self-pollinated. For self-pollinating varieties, such as beans, peas, peppers,

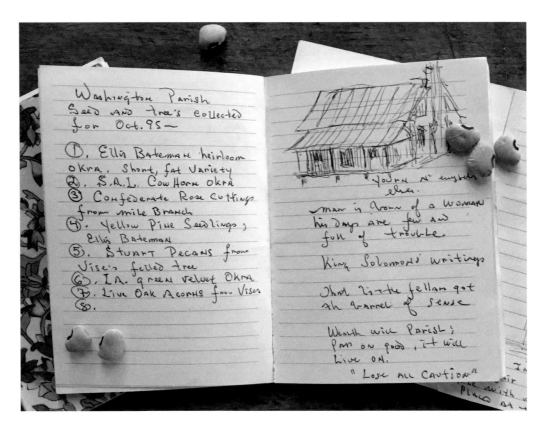

A 1995 journal entry lists seeds I had collected from Washington Parish friends.

greens, and tomatoes, this means that when the plant goes to seed and you plant those seeds, the new plant will be true to type, or the same as the parent plant. However, to make sure that strains remain pure, it is a good idea to isolate the varieties. For recommendations on isolation distances, I recommend consulting Susanne Ashworth's 2002 book, *Seed to Seed: Seed Saving and Growing Techniques for Vegetable Gardeners*.

Many commonly grown vegetables are easily cross-pollinated by insects or by wind. Corn, pumpkins, squash, broccoli, carrots, cabbage, melons, radishes, and turnips are pollinated this way. In the garden, you should therefore isolate each variety, by several hundred yards in many cases. In a small garden, the easiest way to ensure

purity is to grow only one variety of a species at a time. If your goal in raising an heirloom variety is to preserve it, you do not want it to cross with something else.

When it comes to seed saving, I think the "Cadillac" resource for saving endangered seeds is the Seed Savers Exchange. Located in Decorah, Iowa, this nonprofit organization was founded in 1975 by Diane Ott Whealy and Kent Whealy. Diane had a couple of heirlooms that her grandparents had been saving over the years, and Kent was an avid gardener. The then-married couple started their organization with about a dozen members and a small newsletter. Seed Savers Exchange has grown to be the largest seed-saving organization in the world that conserves genetic diversity without government support. The group has over thirteen thousand members and twenty thousand plant varieties. I've been a listed member since 1990.

Every year, the Seed Savers Exchange produces a large catalog that goes out to listed members. The publication contains any growing thing that members want to share, everything from apples to zucchini. Every year, I have an average of ninety listings in that seed book. Anyone interested in a particular thing I list can write me for a starter sample. Some varieties in the catalog are rare. They are listed as "MR," and will only be offered to other listed members and serious growers who "must re-offer" the seeds back after they are grown out. This procedure insures that rare seeds are safeguarded, increased in number, and passed on. Whenever I find a new variety, the first thing I do is to send a sample to the Seed Savers Exchange and then I consider it "in the bank," so to speak. In other words, if I depart this world tomorrow, I won't have to worry about my precious seeds being lost in the freezer.

Even with all these new resources, we are still racing against time to find and save what is left of old seed varieties. Luckily, in the last number of years we've had a lot of dedicated people who've been scouring the countryside across America looking for rare seeds. It is one thing to save these seeds, but unless we find people willing to grow them and increase the seed stock, the work is only half-done and the seeds are not sufficiently protected. One of my favorite things about Washington Parish is the deep tradition of saving seeds and growing gardens. Nearly everywhere you go in the parish, you'll find garden plots and people who take pride in the varieties of vegetables they grow. When a vegetable crop comes in and a family can't use it all, they typically give away the surplus to friends, neighbors, and coworkers.

There is no better gift between friends than freshly grown produce and seeds. My friend Henry Harrison has always had an eye out for unique varieties I have not encountered. He shares rare seeds he finds with me, and I share varieties I come across with him. One of my favorites is a variety of okra his family grows, known as cowhorn okra. Okra itself came to the Americas with the slave trade from Africa, but it quickly made its way onto southern tables in famous dishes like gumbo. The cowhorn variety can grow to an astounding twelve feet in height with pods longer than a foot in length. Despite its size, the plant's pods are known for their tenderness.

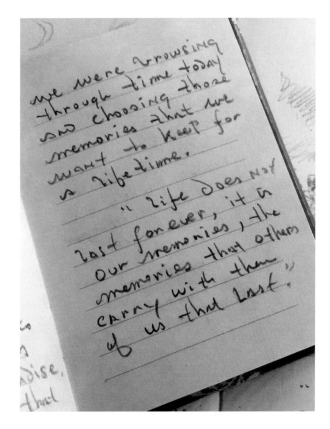

My reflections on memories and the passage of time.

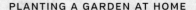

PLANTING A GARDEN AT HOME

For people who want to begin their own home garden, I always recommend starting on a small scale, so you don't overwhelm yourself. It might be a few tomato plants, a row of beans, or a few cucumbers. Learn as much as you can from local gardening experts and backyard gardeners. They often have the inside track on the best varieties and methods for your area. There is a wealth of knowledge out there just waiting to be tapped.

My tips are primarily for the Deep South region of Washington Parish. Growing zones and seasons vary across the country, and there are many resources available for localized information. You can start by looking on the Internet for planting dates and instructions for planting in your area, or by contacting your local farm bureau, agricultural extension agency, or university.

English Peas

English peas are grown for the tender, green-shelled seeds nestled in the pea's pods. They're harvested before the pods toughen and begin to dry out. I plant peas about six weeks before the last frost in loamy or sandy soil that is well-drained. I plant two seeds at a time, about two knuckles deep, and two to three inches apart. I give them some form of support, like wire or string. I like to keep the soil damp. After planting, the peas should be ready to pick in about a month and a half.

Irish Potatoes
In the Deep South, my friends would plant their Irish potatoes by Valentine's Day. For most places, you can plant them about three weeks before the last frost. I take a regular potato and chop it into pieces, making sure there are at least two eyes that can sprout on each piece. I till several inches of soil with compost in an area that drains well. I dig about a four-inch-deep trench, place the seed potatoes in the trench about a foot apart, and cover the potatoes with soil. When I want to harvest them as "new" or immature potatoes, I usually pull them up about two and a half months after planting, before the vines die, about ninety-five to a hundred days after planting. Mature potatoes are usually ready in about three months.

Lettuce
Head lettuce is a cooler-weather crop. I plant it in early spring before the summer heat sets in. Lettuce seeds germinate under a very thin layer of soil. You can either scatter seeds in a bed or plant seeds about a foot apart in rows. Baby lettuce can be harvested within two weeks after germination. Just cut the whole plant right above the ground, once it is around three to four inches tall. Mature head lettuce can be harvested in about sixty to seventy days from planting.

Beans, Butter Beans, and Field Peas
Generally, if you are growing heirloom beans, butter beans, or peas, there are two kinds, climbing and bush. The process of planting and growing them is pretty much the same. Generations ago, farmers planted field peas in their corn patch, where they would use the tall corn plants as support for the peas to climb. Today, we use anything for support, including string or wire fencing.

In Washington Parish and other areas of the Deep South you can start planting field peas and beans in April, or as soon as the threat of frost has passed. I usually soak my seeds the night before planting. I plant heirloom varieties about two to four bean or pea seeds to a hill, about knuckle deep, on raised rows spaced four to five feet apart. They need this much space

because heirloom varieties tend to "run," meaning their vines creep. Modern types don't run as much as heirloom varieties, and you can get away with planting them three or four feet apart. If planting in dry season, I always carry a bucket of water to pour over the planted seeds. Heirloom varieties are pretty self-sufficient and don't need regular watering. To get enough beans or peas for a family meal, you'll need a good-sized patch, probably at least two ten-foot rows. A pea plant has various stages of growth on it at the same time, and the beans and peas are edible at every stage. My favorite varieties are Reverend Taylor Butter Beans, Mr. Seldon's Red Ripper Cowpeas, the Unknown Pea of Washington Parish, and clay peas.

Tomatoes

Good homegrown tomatoes actually start with preparing the soil about three months before you plant. I always recommend having your soil's pH tested, or testing it yourself. Then apply the amount of lime recommended by your local extension service. Adding lime to soil boosts calcium and other nutrients that help prevent common tomato diseases such as blossom-end rot. Just remember, lime needs two to three months to break down in the soil before planting.

Tomatoes are a warm-weather crop. I wait to plant until the danger of frost has passed and the soil has warmed. When I grow an heirloom variety seed, I start seeds in a greenhouse environment about six to eight weeks before planting time. Once the plants are about a foot tall, they are ready to put in the ground. I like to space them about four feet apart so air can flow in between each plant, which prevents some diseases. I water tomato plants when I plant them, and I keep them consistently moist, not dry or heavily watered or overwatered. A good test for moisture is to put your hand down and feel the soil. If it's dry, water it. I don't usually prune my tomatoes much, and I keep them at about four main stems. My favorite heirloom variety is the Gulf State Tomato, which was first introduced in 1921.

Okra

Okra is a favorite vegetable in the Deep South, and it's another warm-weather crop. Okra seeds are hard to sprout. If you put them in dry soil, they can take weeks to germinate. My trick is to estimate how much seed I need, pour my seeds onto a paper towel, and wrap them up. I soak the bundle in water; then seal it in a plastic zip-storage bag. I keep the bag in a warm location, checking the seeds every two to three days, until they sprout. When the sprouted seeds are ready to go in the raised rows, I soak the ground with a hose and plant the seedlings about knuckle deep, a little more than a foot apart, and cover them with dry dirt, to keep the surface loose. They'll be ready to pick in about a month and a half, but don't let the okra pods sit on the stems for long. Tender is best.

Spring Onions

Many people know green or spring onions as the long, thin green-top onions, shallots, or scallions found in the grocery store. Although called spring onions, they are planted in the fall. The green shoots come up in the spring. Left to grow, some bulbs can mature into full onions, depending on the variety you plant. My friends in Washington Parish primarily grow what are known as multiplier onions. These grow into clumps of smaller onions and are usually used for the green top. Around June, you dig those up and separate them and plant them again in the fall. For larger yellow, white, and red onions to mature, you have to pull the soil away from the bulb and leave them sitting on top of the soil. If you do this, as my friend Mr. Calvin Vise taught me, "They'll blow up like a balloon."

One of the greatest lessons I have learned in the gardens and at the tables of my Washington Parish friends is that serving meals using fruits and vegetables fresh from the garden is simple, and requires few ingredients and little time. Eating this way is delicious and healthy, and gives you time to enjoy the people around your table. Many things don't even have to be cooked at all. Farm families in days past knew this and lived by it. They didn't have time to spend fretting over meals.

Over the past half-century, we've been sold an idea that pre-prepared "convenience foods" are the way to go. Lately, television shows have gone from teaching us how easy and improvisational cooking can be to turning mealtime into a precise science experiment or a cutthroat spectator sport. What I have found is that cooking fresh, frozen, or dried vegetables in simple ways leaves me with a satisfying, tasty meal and plenty of time to celebrate friends and family at the table. To me, those are the memories that stick.

SUMMER

Summer heat in the Deep South is something you wear. Its sticky heaviness is so oppressive, it takes your breath away. In the days before electricity and fans or "chilled air," muggy heat was inescapable, but that didn't stop hardworking farm families from doing their daily work. To avoid the worst heat of the day, they rose before the sun to start hoeing or "chopping" weeds out of corn and cotton patches, and to make sure cash crops were coming up nicely. They occasionally paused from their work to cool off with refreshing well water or a stray melon strategically planted in corn or cotton fields for a special treat. They prayed for the rains that they depended on to water crops. On special occasions, when their work was done they cooled down in nearby streams and swimming holes.

Summer was the time to collect the garden's bounty and the fruits of their labor, so time spent in the fields had extra reward and urgency. As soon as summer vegetables started coming in, women and children were hard at work gathering them from the garden. Every day, they picked, pulled, shelled, and put up what they grew.

Ripe vegetables were best eaten the day they were picked. Mothers would often send their small children out into the garden to gather just enough vegetables for dinner or supper. Leftovers made their way back to the table for the next meal, even for breakfast. The rest was canned, pickled, and dried for later use. On hot days, summer fruits like watermelons, peaches, figs, blackberries, scuppernongs, a variety of muscadine grapes, and cultivated berries such as strawberries and blueberries

Pink tomatoes.

123

made refreshing snacks. These fruits also made delicious pies, jams, jellies, preserves, and even pickles, for year-round summer flavor.

FIG PRESERVES

Makes 10 pints. Adapted from a recipe from the Mile Branch Settlement Cookbook Committee.

———◇———

I am amazed by the number of Louisianans who have enormous, old fig trees. Once figs come in during the summer, preserves are the only way to capture their deliciousness to enjoy year-round.

10 pounds fresh figs, stems removed

4 cups sugar

1 medium lemon, sliced

1. In a large container, pour sugar over figs and let sit overnight in the refrigerator.
2. The next day, sterilize 10 pint-sized glass jars and lids. Transfer fig mixture and lemon slices to a large pot and cook over low heat until fruit syrup is thick, about 3 hours. Stir every once in a while.
3. Pour hot preserves into sterilized jars and seal according to manufacturer's directions.

I often write down charming things my friends say to me, like these comments from a summer visit in 2002.

Facing page: Washington Parish is well known throughout Louisiana for its prized watermelons. The acidic, sandy soil contributes to their sweet flavor. For years, I have been fascinated by all of the varieties grown in this area, and I have sampled as many as I could find.

I rarely spent summers in Louisiana, although I was always imagining what the gardens tended by my friends looked like at that time of year. When it was time to plant my own yard in Tennessee, I would often use their seeds to pay homage to their efforts and to increase the stock of their varieties. I detailed one of these gardens in an August 1991 journal entry:

Some of my best seed savers projects are taking place right in the front yard. As far as the looks of the garden go, this might be a garden sometime during the 1880s. A row of "Snow on the Mountain" pole limas climb tall cane poles and spill over the sides, reaching heights of over twenty feet. Close by and spreading out over the yard are long vines of Tennessee sweet potato pumpkins. These light cream-colored, brightly striped with green and bell shape, stand out among the vines. On the other side of the limas are the old varieties of large pink tomatoes. Already harvested and drying are the multiplier potato onions which are another variety common at that time. This is a summer of living heirlooms at a time of growing with and walking among examples of living history. Working with the beautiful varieties reawaken vivid pictures of times where these varieties were first grown.

From my home in Tennessee, I still regularly speak with friends in Washington Parish to get reports on how things are coming along in their gardens. Even today, the Vise and Lang families observe the tradition of getting together to pick, put up, and divide up their fruits and vegetables, packing freezers full. The summer bounty ripens so quickly. They often find themselves picking every few days and inviting friends, neighbors, and coworkers to help themselves to whatever they can carry home.

Even though farm families were grateful to have food on the table, the regular staples could wear out their welcome. I vividly recall one old-timer telling me that, when he was little, one day he had just had enough of eating peas. He sat at the table and pushed his bowl of peas away, telling his daddy, "Daddy, I'm sick of peas and I ain't eatin' no more peas, I'm tard [tired] of 'em." His daddy, who was a kind and understanding man, simply replied, "That's all right son. You'll eat 'em tomorrow."

The produce farm family cooks had to work with didn't change much, but that did not discourage them from being creative and spontaneous in the kitchen. As in days past, some of the best meals are made of garden vegetables served as simply as possible, with the addition of a few seasoning ingredients. New potatoes, okra, corn, and summer squash can be roasted, sautéed, stewed, fried, or steamed with meat fat, butter or olive oil, and salt, pepper, and a few spices. Tomatoes, okra, onions, and garlic can be sautéed together or stewed with seasoning meat. Young, tender ears of corn, picked in what's called the "milk" stage, can be roasted in their husks in the coals of the hearth fire or woodstove. Or it can be cut from the cob and fried in a cast-iron skillet or "creamed," which involves just what you think it does, cooking fresh corn kernels in rich milk, a little flour, and butter. Best of all, fresh summer vegetables are delicious all by themselves. Tomatoes, cucumbers, onions, and radishes can be cut up and tossed with a little vinegar or mayonnaise, garlic, salt, pepper, or spices.

Facing page: Summer is the busiest time of year for my Tennessee gardens. I often plant the seeds I gathered from my many trips to Washington Parish, in order to increase seed stock, but also to honor my friends who shared the seeds with me. They taste good too!

Many of my friends and their families still pick and put up summer crops. After I grew out my stock of the Unknown Pea, I gave some to Mr. Seldon's son and my good friend Mike Lang. He planted a field of them in 2016. Christina Melton, who helped me write this book, along with Sarah Hackenberg, who took the photographs for the book, brought their families to Lang's farm to help them pick and put up the first crop. Mike Lang's grandson even helped pick. It was an incredible example of carrying seeds and traditions through and between family and friends. It brought back a memory of one of the last phone calls I had with Mr. Seldon. He told me, "We got six five-gallon buckets on the first pickin'. I planted them in hundred-foot rows. We used that pea sheller and shelled out two big dishpans full of 'em. I shelled out a sack full of dried peas for seed. The red speckled butter beans you gave me made good too, I had a row in the patch beside the house. I had around eighty tomato plants. Sold a few to my regular customers. When you get down here we'll get in the shade and talk about it. Tater pumpkins are making." On working in the fields he remarked, "I give out quick."

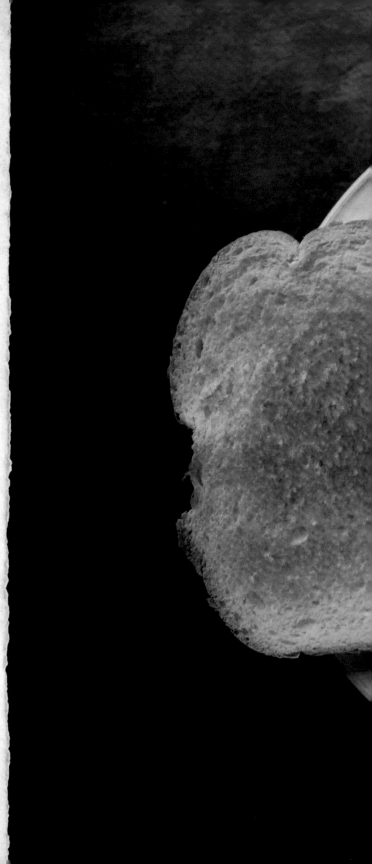

Tomato sandwich with
homemade mayonnaise.

HOMEMADE MAYONNAISE

Makes 1 cup. Courtesy of Francis Adcock
of Monroe, Louisiana.

———◇———

There is something about homemade mayonnaise that makes anything taste good. This recipe has been handed down in my friend Christina Melton's husband's family for several generations, and it's delicious on fresh tomatoes.

2 fresh egg yolks
½ teaspoon salt
½ teaspoon dry mustard
½ teaspoon sugar (if desired)
1 tablespoon white vinegar
1 teaspoon warm water
1 cup vegetable oil
1 teaspoon lemon juice
A dash or less of cayenne pepper

1. Place egg yolks in a medium bowl and mix in salt, mustard, and sugar. Stir in vinegar and water to make a paste.
2. Whisk mixture continuously as you slowly dribble in the oil. As mayonnaise thickens, continue to whisk and slowly add lemon juice and red pepper. Keep mayonnaise in a sterilized, covered jar in the refrigerator.

MAW LANG'S ENGLISH PEA AND TOMATO SALAD

Makes 6 servings. Courtesy of Bob Ann Breland.

———◇———

This is a quick salad Maw Lang made over the years, and it continues to be a favorite of her family.

1 medium tomato
2 cups tender, fresh English peas, shelled and blanched
Spoonful of mayonnaise, preferably made fresh; see preceding recipe
Salt and pepper to taste

Dice the tomato. Put the tomato and peas in a bowl and chill until cool. Stir in mayonnaise, salt, and pepper and serve.

SUMMER SUCCOTASH

Makes 6 servings. Recipe adapted from my journal notes.

———◇———

This dish is a familiar southern staple on summer tables. It is made from a base of butter beans, corn, and tomatoes. But really, you can add whatever fresh garden vegetables you have on hand.

2 cups fresh or frozen butter beans
½ small yellow onion, unchopped, plus 1 small yellow or sweet onion, chopped
1 garlic clove, crushed
4 ears sweet corn
3 uncooked bacon slices
1 cup sliced okra (optional)
1 cup chopped red bell pepper (optional)
1½ cups cherry tomatoes, cut in quarters
2 tablespoons unsalted butter
1½ tablespoons chopped parsley
Salt and pepper to taste

1. Place butter beans, small onion half, and garlic in a pot and cover with water. Bring to a boil and reduce to medium high heat. Simmer beans, uncovered, until tender, about 15–20 minutes. Drain beans, reserving ¾ cup cooking liquid. Discard onion and garlic.

Just-picked okra.

2. Cut corn from cobs and scrape corn milk and set aside. In a large skillet, cook bacon over medium heat until crisp. Drain bacon on a paper towel and leave 3 tablespoons of the drippings in the skillet.

3. Over medium-high heat, add chopped onion to drippings and sauté until tender. (If using chopped okra and red bell pepper, sauté them with the onion.)

4. Stir in corn and corn milk and cook over medium heat until corn is tender. Stir in tomatoes, cooked butter beans, and reserved ¾ cup cooking liquid. Cook, stirring occasionally, for 5 minutes. Don't let beans get too mushy or the tomatoes overcook.

5. Stir in butter, parsley, salt, and pepper and toss with crumbled bacon. Serve warm.

FRIED OKRA

Makes 4 servings. Courtesy of Bob Ann Breland.

My friends make fried okra in a variety of ways. Mrs. Luella Crane used to just slice it up and toss it in a pan of bacon grease, sauté it until it was tender, and finish it off in the oven until it crisped up. That's so simple you don't need a recipe. When most people think of fried okra, they think of it crusted with cornmeal, and the following is that particular recipe from my friend Bob Ann.

Peanut oil for frying (enough to cover 2 inches
  deep in your pan)
1 cup buttermilk
1 large egg
4 cups tender, fresh okra pods, cut into ⅓-inch slices
1 cup finely ground cornmeal
1 teaspoon salt
½ teaspoon cayenne pepper

1. Heat oil to 350°F in a deep, heavy-bottomed skillet or a Dutch oven. In a large bowl, whisk together buttermilk and egg. Drop in the okra slices.

2. In a separate bowl, mix together cornmeal, salt, and cayenne pepper. Pull a handful of okra slices from the buttermilk mixture and drain. Drop drained okra in cornmeal mixture and toss to coat.

3. In small batches, drop coated okra in hot oil. Fry, turning often, until light brown and crispy. Drain and cool slightly. Serve warm.

SKILLET CORN

Makes 6 servings.

Recipe adapted from my journal notes.

———— ◇ ————

8 ears corn

10 slices bacon

1 small yellow onion, finely chopped

3 tablespoons half-and-half

Black pepper to taste

1. Cut corn from cob and scrape corn milk. Set aside.

2. In a large, heavy skillet, cook bacon over medium heat until crisp. Drain bacon on paper towels and reserve 2 tablespoons of the drippings in the skillet. Sauté onion in reserved bacon drippings until tender and brown. Add cut corn and corn milk and turn up the heat. Cook corn until it browns, stirring occasionally.

3. Reduce heat and stir in half-and-half and pepper. Bring to a simmer and remove from heat. Serve warm.

On a few occasions, I was able to travel to Washington Parish during the summertime. These were special trips for me because I was able to help with harvesting. I also saw how my friends used what they grew, from serving vegetables straight from the fields, to pickling and canning as they have for generations. Mr. Arlie recalled his mother stoking fires under enormous cast-iron vats of boiling water used to sterilize glass canning jars, while stirring other large pots filled with fruits and vegetables ready to be put up. His mother would wrap the jars in towels so they wouldn't break in the roiling water. "She'd be up way past midnight, until one or two in the mornin' fillin' and sealin' those jars with beans, tomatoes, okra, and all sech es that." I always say the best instructions for canning are available from canning companies, who know what they are talking about. I follow the instructions on the side of the box of jars, or I look it up online.

REFRIGERATOR DILL PICKLES
Makes 3 pints.

—◇—

I always like the simplest form of pickles made from fresh cucumbers, salt, and vinegar and kept in the refrigerator. You can use this recipe with green tomatoes, okra, and green beans, too. Fresh vegetables prepared this way keep about two weeks.

- 4–5 large, fresh cucumbers
- 2 cups water
- 1 cup white vinegar
- 1 tablespoon kosher salt
- 1 teaspoon sugar
- 3 cloves garlic, peeled and crushed
- ½ teaspoon dried dill seed, or one bunch fresh dill sprigs
- ½ teaspoon dried mustard seed
- ¼ teaspoon red pepper flakes, or to taste

1. Sterilize 3 pint-sized glass jars and lids. Wash the cucumbers well and cut into ¼-inch thick slices.
2. In a large saucepan, bring to a boil the water, vinegar, salt, and sugar and cook until solids dissolve and the liquid is clear, about 5 minutes. Add garlic, dill, mustard, and red pepper flakes.
3. Place cucumbers in the jars, but do not pack tightly. Pour hot pickling brine mixture over the cucumbers. Allow to cool, and then screw on the lids. Store in the refrigerator up to 2 weeks.

Often I'm able to capture simple tips and recipes for preserving the summer's bounty, like this 1990 journal entry about canning peppers, cucumbers, and beets.

SAVING SEEDS

When people ask me the secret to saving seeds, I tell them it is simple. I think seed saving is one of the most basic things anyone can do. Our ancestors were seed savers by necessity. They didn't have a seed store to go to, and they didn't have money to purchase seed. It was crucial for a family's survival to save their seeds year after year. The techniques they used to harvest and preserve seeds were passed down with the seeds.

Today, storing seeds is easier than ever before, and we can all play a role in saving and sharing them. I always say the three main enemies of seed are heat, light, and moisture. You should never dry seeds in direct sunlight or store them in damp conditions. When storing properly dried seeds, especially beans, peas, and corn, plastic containers and zipper bags protect well in the freezer. Envelopes and clean plastic medicine vials work well for smaller seeds, such as tomato, okra, and squash.

I try to encourage people to become self-sufficient when growing their own food, even if they use their own seeds for just one tomato or one bean variety. It's gratifying to know that what you harvest is something that's yours, and that you were totally independent in growing it. To save vegetable-seed varieties, here are a few basic tips.

Beans and Peas

English peas, field peas, butter beans, and most varieties of beans are probably the easiest seeds to save. To make sure beans are fully mature, I most often let the pods dry on the vine. After I collect the dried pods with the peas still inside, I sit on the porch or by the fire and shell them. Many of my Washington Parish friends put the hard, dried pods in a burlap sack or a crawfish sack and beat them with a stick to break open the pods and let the peas and beans fall out. I usually store seeds for the next season in the freezer in a tightly sealed container or jar or in a plastic bag. I label the bag with the seed name and harvest year. Freezing keep pests away. Dried hulls are good for kindling.

Okra

Saving okra seeds is a process similar to saving other dried pods. I usually allow okra pods to dry on the stem or away from heat and light. I then break open the dried pods and separate the seeds. Okra seeds should be stored in a cool, dry location.

Melons, Pumpkins, Squash, Peppers, and Eggplant

Saving seeds from these vegetables is a little different because you harvest seeds from the ripe vegetables. Once you remove the seeds, rinse them thoroughly in cold water. Spread the seeds out on a nonstick surface (like a cookie sheet) to dry, away from direct sunlight. Drying usually takes about a week. Once dry, store the seeds in a paper envelope or tightly sealed jar. I often use clean, recycled prescription-pill vials.

Tomatoes and Cucumbers

Saving tomato and cucumber seeds is slightly more complicated than most seeds, but not difficult. First you have to let the tomato ferment to remove the gelatinous membrane that surrounds the seed. To do this, remove the seeds from the fruit and let them sit in a cup, out of direct sunlight, until a whitish mold forms on top, usually about three to four days. Rinse the seeds thoroughly in a fine sieve to remove the slimy goo. Then spread them out on a nonstick surface to dry, away from direct sunlight, for about a week. Store them in a paper envelope, tightly sealed jar, or prescription-pill vial.

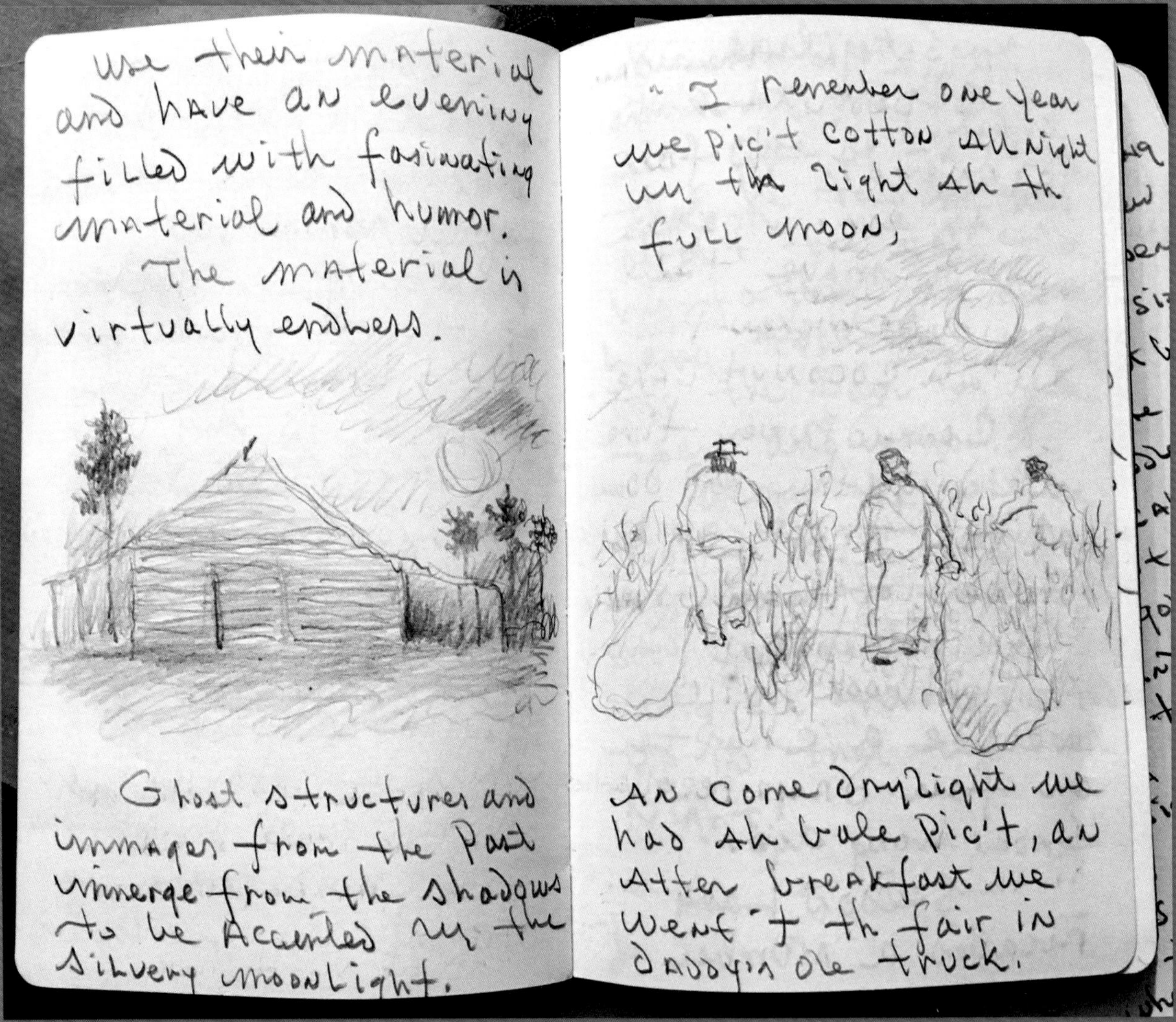

use their material
and have an evening
filled with fasinating
material and humor.
The material is
virtually endless.

Ghost structures and
immages from the Past
immerge from the shadows
to be Accented by the
silvery moonlight.

" I remember one year
we pic't cotton all night
un tha light ah th
full moon,

An come daylight we
had ah bale pic't, an
After breakfast we
went t th fair in
daddy's ole truck.

A 2010 sketch and journal entry recording Homer Graves's memory of picking cotton by moonlight.

The hottest part of the summer was when cotton and corn, the most valuable cash crops on many Washington Parish farms, started to come in. Homer Graves remembered scraping weeds out of as many as twelve rows of cotton a day: "It went pretty fast. Sometimes you would have to pull weeds from around plants, and then in other parts of the row there wouldn't be much at all to scrape." Mr. Homer also remembered the Fourth of July as more than just a celebration of the nation's birthday. "Seems like we always poisoned cotton on the Fourth of July," he said, recalling putting out poison to ward off boll weevils, a scourge that nearly decimated southern cotton crops during the first part of the twentieth century. "We used old whiskey bottles with holes cut in the cork tops to shake the poison out. We went down those rows shaking the bottles over the cotton bolls. Daddy told us we could go to the Fourth of July celebrations as soon as we finished."

Growing cotton was not easy. Not only did growers face boll weevils, but increasing mechanization at the turn of the century meant farmers who still tilled the earth with mules had a hard time keeping up. During the Great Depression, prices for cotton plummeted. The New Deal's Agricultural Adjustment Act implemented in the 1930s helped small farmers, but it often hurt sharecroppers, because only landowners were eligible for assistance. Sharecroppers had few other ways to make a living and to feed themselves, and they were hit particularly hard. Mr. Homer recalled government attempts to control cotton prices, which meant limiting acreage. He told me a story about how one surveyor told his father he had planted too much cotton and would have to plow the excess under. "Daddy told that man, 'The hell you say. I planted that cotton. We're gonna pick it and sell it.' And that's what we done."

Late in summer and into the fall, another staple crop, sugarcane, went into the ground. Few farmers in this piney woods region grew it on a large scale, but most had a small patch of their own because they could not afford white sugar. Cane syrup and cane sugar were essential staples for farm families for cooking. Cane stalks, harvested the year before, were carefully laid in trenches in the ground. New sprouts formed out of the joints in the cane stalks. Sugarcane would be harvested in the early winter around the first frost, when the sugar content was high, and it was processed into cane syrup that lasted throughout the remainder of the year. Surplus syrup was used to barter or sell for money to buy things farmers couldn't make or produce themselves like salt, coffee, or tools.

The daily rituals of working together in the fields, gardens, barns, and kitchens were widely observed in farming families up until the middle of the twentieth century. But many other rituals brought the entire community together, beyond family bonds. These were spiritual rituals that spanned the lives of the community members, from birth to death.

Since the late 1800s, church revivals and other religious gatherings were common in late summer and early fall. Families often traveled rough roads and narrow pathways by horseback and mule-drawn wagons and buggies to attend weeks-long tent revivals that were full of rousing sermons, singing, and communal meals. Baptisms, along with other religious rituals, were regularly held in nearby creeks and along riverbeds. Church congregations would get together for old-time "foot washings," where members would follow Jesus's example of washing someone else's feet as a sign of humble service to one another.

When people weren't traveling to revivals, they at-

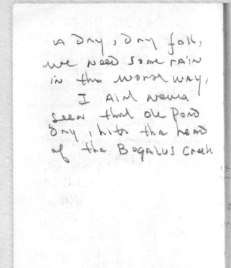

An October 11, 2002, journal entry about
the need for rain.

Rain was welcome during
the hot summer to freshen fields, as
described in a 2002 journal entry:

After last night's summer rains, the
morning air is alive with life. You can
smell the fresh green growth soaking
up the drenched fields. I had almost
forgotten the sensation of a Deep
South's early morning hours, where
the humidity feels so dense that you
feel it running off your forehead before you reach the field.

tended their own community church services at least twice each week, especially on Sundays. These sermons often lasted a large part of the day. Mr. Seldon told me a story about a certain Sunday sermon that ran a little long. The preacher asked how many among the congregation wanted to make the trip up to heaven. Mr. Seldon recalled, "Everyone raised their hand with the exception of one old man who was sitting on the front bench. When the preacher inquired as to why the man had not raised his hand, the man replied, 'I thought you was getting up a bunch to go today and I ain't ready to go. I still got corn to pull and cotton to pick.'"

When preachers were not busy with Sunday sermons or staging revivals, they were traveling dusty back roads spreading the gospel. Preachers often carried seeds to hand out to the people they ministered to. These seeds varieties often became known by names such as the "Preacher Pea."

One long-held tradition in this region is the annual homecoming. Homecomings are gatherings on church grounds where generations of church members from near and far come back together for a spiritual renewal and fellowship. I have been fortunate to attend many homecomings at the Carson Springs Baptist Church, which is located in the Little Improve Community up in Mississippi, not far across the state line from Washington Parish. I once saw a sign there which read, "Welcome to the Little Improve Community, conveniently located in the middle of nowhere."

The service at the Carson Springs Baptist Church begins with an inspiring sermon, along with the singing of popular traditional hymns. As in times past, these homecomings always end with "dinner on the grounds." Long before air-conditioning, long, wide plank tables were set outdoors on the church grounds in the shade of tall pines. The tables were filled with dishes brought to share, and there was enough food to feed everyone before the long journey home. Today, this homecoming meal is held in the church hall.

I look forward to the welcome I receive at these homecoming services and to the friendly, familiar faces I see year after year. I am always impressed by the meal that follows, with each family bringing its own favorite recipes. The individual talents of the cooks and their many years of experience are evident in each variation of a basic dish. There are numerous chicken pot pies with heavenly crusts. Then there are several offerings of chicken and dumplings, with no two alike. Further down the line are many examples of beef roast and gravy. One lady made such an outstanding version of roast with gravy that I can still taste it today. There are also platters of fried chicken, pork roasts, bowls of creamed corn, and turnip greens with roots. There are collards, crowder peas, butter beans, white beans, green beans, candied sweet potatoes, mashed potatoes, potato salad, and many dishes I can't remember.

To me, chicken and dumplings is the epitome of southern comfort food. In my visits to this part of the country, I have eaten more renditions of this dish than I could ever count, but the Lang family recipe is my favor-

There were what they called Pew dolls; they were homemade and they gave them to the children so that they would be entertained and not cause a fuss in church.

A 2008 sketch and journal entry describing "pew dolls" that young children would play with during church services to help keep them quiet.

ite. Bob Ann Lang Breland and her sister-in-law, Daisy Lang, told me that a meal of chicken and dumplings or its close relative, chicken pie, was reserved for special occasions. Both are delectable concoctions made with chicken, flour, shortening, butter, and milk. Bob Ann explained to me that the difference between chicken pie and chicken and dumplings is that chicken and dumplings are cooked in the pot until done and usually served over rice. Chicken pie is baked in a baking dish and topped off with a "blanket" of golden crust. Bob Ann also told me that, when they were growing up, nobody owned a roasting pan. When her mother needed a pan large enough for a big chicken pie, say for the fall revival at church, the biggest container for baking in the house was a dishpan. She recalled that the way you knew whose chicken pie you were eating was by the dishpan it was served in.

LANG FAMILY
CHICKEN AND DUMPLINGS

**Makes 6 servings. Adapted from recipes from
Daisy Lang and Bob Ann Breland.**

———◇———

To make this dish, Beulah Mae Lang would sterilize her dishpan, which she would put on the stove with the chicken and salted water. She boiled the chicken in the dishpan until it was tender, then dropped in the rolled-out dumplings. Next, she added golden butter made from the cream from the family cow. Bob Ann made the butter herself, by shaking cream in a quart jar until the golden drops of butter separated from the buttermilk. (Bob Ann's butter recipe is on page 63.) When the dumplings were dropped into the boiling pot of chicken, Bob Ann distinctly remembers the sound of her mother shaking the pan to keep the dumplings from sticking together, and the smell of something beyond wonderful. Of course, if you don't have a large dishpan, you can use a large sauce pot to make this recipe.

1 (2½ to 3 pound) chicken

2 quarts water

2 teaspoons salt

1 teaspoon ground black pepper

Dumplings (recipe follows)

4 tablespoons butter

4 tablespoons heavy cream

1. Boil chicken in the water with the salt and pepper until it is tender, about an hour. Remove chicken from broth, reserving broth. When chicken is cool, remove meat from bones. Discard skin and bones and add chicken back to broth. Let it sit while you mix together and roll out dumplings.

2. Reheat broth and chicken mixture. Add butter and cream to broth, and let it sit at a simmer.

3. With a knife or pizza cutter, slice paper-thin, rolled-out dumpling dough into rectangle-shaped pieces about 1 inch wide and 2 inches long. Drop dumplings into simmering broth mixture. Continue rolling out small balls of dough, cutting and adding to the broth, and stirring gently until dumplings are pretty thick in the broth. Be sure there's enough liquid broth to cook the dumplings without them sticking together. Add extra both as needed.

4. Continue stirring gently until all dumplings are cooked and stew thickens, about 20 minutes. Serve hot in bowls.

DUMPLINGS

Makes 6 servings.

———◇———

2 cups all-purpose flour

½ teaspoon baking soda

½ teaspoon salt

3 tablespoons vegetable shortening or lard

1 cup buttermilk or water

1. Sift flour, baking soda, and salt into a large bowl. Make a well in the center and add shortening, then buttermilk and mix with a spoon. When dough gets thick, sprinkle with a little flour and knead in the bowl by hand until it is smooth and tough. (This step is important.)

2. To roll out dough, sprinkle a hard surface with flour. Rub a rolling pin with flour, and have plenty of flour on your hands. Pinch off small balls of dough about the size of your palm. Sprinkle it with flour and roll paper thin. Cut dumplings according to your recipe's directions.

CHICKEN PIE

Makes 6 servings.

———◇———

To make chicken pie, pour the hot chicken-and-dumpling mixture from the Lang Family Chicken and Dumplings recipe on page 142 into a 13×9-inch baking pan. Prepare an additional recipe of dumplings. Roll that dough flat and lay it on top of the pan. Brush the dough with melted butter and put the pan in a 350°F oven and bake until the crust is brown. This should take a little over a half-hour.

Just at the point when I feel I can't hold one more bite, I always find more tables filled with desserts. Among the many choices are two of my favorites, sweet potato pie and pound cake. Over the years I have never made it past these two confections. I once told someone that on my next visit I plan to start with dessert so I can sample all of them.

COCONUT POUND CAKE

Makes 10 to 12 servings.

Courtesy of Bob Ann Breland.

—◇—

This recipe from Maw Lang is a true family tradition. I have had this pound cake or similar renditions at many Lang family meals and celebrations. It is delicious for dessert, and also as a snack with coffee or milk.

3 cups sugar

¾ pound (3 sticks) butter

6 large eggs

3 cups all-purpose flour

¼ teaspoon baking soda

8 ounces sour cream

¾ cup shredded coconut (Maw Lang used Baker's Angel Flake)

1 teaspoon coconut or vanilla extract

Cake glaze (recipe follows)

1. Bring all ingredients to room temperature. Preheat oven to 325°F. Generously grease and flour a 10-inch tube pan. With an electric mixer, cream together sugar and butter for 6 minutes. Add eggs one at a time, beating well after each addition.

2. Sift in flour and soda and mix well. Stir in sour cream, coconut, and coconut extract.

3. Spoon batter into prepared pan and bake until a knife inserted in the center comes out clean, about 1½ hours. Remove from oven and cool 5 minutes. Flip pan over and place warm cake on a platter. Top with glaze and serve.

CAKE GLAZE

Makes about 1 cup.

—◇—

¾ cup sugar

¾ cup water

1 teaspoon almond or vanilla extract

Bring sugar and water to a boil and cook until sugar is completely dissolved. Add extract and pour over warm cake.

On most Sundays, families looked forward to dinner at home after services. In early days, it was customary to invite the preacher and his family to join them. Preachers gladly accepted invitations, as they depended on the goodwill of parishioners to supplement their meager earnings and to help feed their families. Mr. Seldon's mother would go to great lengths to prepare the best she had to offer, fried chicken, which was Mr. Seldon's favorite meal. He vividly recalled the heavenly aroma of chicken frying in the kitchen while the children milled about on the front porch and played in the front yard. He told me that was the one memory that still made his mouth water.

Mr. Seldon said the dining habits for children and family were quite different from what we are used to today. In those days, the adults ate first and the children waited their turn, in hopeful anticipation that there would be something left of the main course. Usually, the

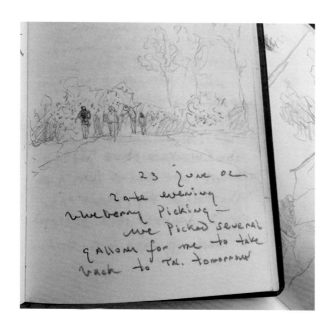

A June 23, 2002, sketch of my friends Calvin and Rose Vise blueberry picking on their farm.

children had to settle for "sides." Mr. Seldon recalled that one particular Sunday the preacher and his wife visited the Lang home for dinner. While the men exchanged pleasantries and news on the front porch, the women made their tour of the garden. The wait was agonizing for Mr. Seldon. Finally, his father, Mr. Albert, took out his "dollar pocketwatch" and glanced at the time. It was noon, time for the meal. As the adults seated themselves around the table, the children wasted no time taking up watchful positions, peering in through the windows and screened doors. The young Seldon hopefully looked through a crack in the board-and-batten wall, praying there would be some fried chicken left for him.

As the children hungrily watched, the contents of that platter disappeared rapidly until, at last, there was but a single piece of fried chicken remaining. How would the children divide up this one piece? Unfortu-

nately, their dilemma was solved when the preacher helped himself to the last serving. Mr. Seldon laughed that he was a grown man before he knew that there was anything more to a chicken than the bony back and the neck. But at least there was plenty dessert, most often some delicious pound cake, fruit pie, or cobbler.

ROSE VISE'S BLUEBERRY PEACH COBBLER

Makes 10 servings. Courtesy of Calvin Vise.

There is no better use for fresh summer fruit than in a cobbler. This particular recipe can be made with blueberries or peaches, or, best of all, a combination of both.

3 pints blueberries

2 teaspoons freshly squeezed lemon juice

2 pounds peaches, sliced

¼ cup brown sugar

¾ cup (1½ sticks) cold unsalted butter

1½ cups all-purpose flour

¾ cups sugar

2 teaspoons baking powder

¾ teaspoon salt

½ cup buttermilk

1 large egg

1. Preheat oven to 400°F. In a bowl, combine blueberries with lemon juice and set aside. In a separate bowl, sprinkle peach slices with brown sugar and set aside.

2. To make cobbler dough, slice butter into ¼-inch pieces and place it in a large bowl. Add the flour, sugar, baking powder, and salt. Using your hands or a pastry knife, work the ingredients until the mixture resembles coarse cornmeal. Make a well in the middle of the flour-and-butter mixture and add buttermilk and the egg. Stir to combine.

3. In a 9×13-inch baking pan, spread a layer of half of the blueberries. Cover with all of the peaches. Use a rubber spatula to top with remaining blueberries.

4. Dab the cobbler dough on top, smoothing it as much as possible. Bake until the fruit is bubbly and the crust is golden brown, about 40 minutes. Cool slightly and serve.

The daily life of my friends revolved around work, family, and faith, and it was interwoven with other community traditions and rituals honoring and celebrating the natural course of life and death. Mr. Homer told me how the community came together when someone passed away. He said the church bell would toll loudly for every year the deceased lived. "We'd count the number of tolls of the bell and my daddy would know who had passed. If the bell rung once we'd know it was a baby. My ole daddy built many a casket out of that old yellow pine," Mr. Homer told me. "When somebody died back in them days, you had to get them in the ground pretty quick, especially in the summer days when it was hot. My daddy would gather us up with picks and shovels on the ground slide and we would head off for the graveyard. We could get up to a dozen men in a hurry for a grave diggin.'"

As Mr. Arlie recalled,

The night before the burial, family members and neighbors would bathe the body. They'd lay 'em out on a door that was covered with a quilt an' set on sawhorses and they'd take turns settin' up with the dead through the night. The womenfolk lined the casket with cotton and they made a cotton-filled pillow to rest the head on. Come mornin', they'd load the casket on the wagon and head for the graveyard. Family and neighbors gathered at the gravesite and the preacher'd say a few words. We'd lower the casket into the grave using ropes and then we'd go to shoveling the dirt in. That's the way it was back in them days.

Community members would prepare comforting meals that the family of the deceased could eat in their grief and serve to visiting family members and mourners who traveled long distances to pay their respects.

The arrival of a new baby was also a time for the community to come together. It was rare for rural communities such as Washington Parish to have a doctor, so most families relied on experienced midwives, who were often called on in the middle of the night to aid if a birth was difficult. After a delivery, women in the community took turns caring for new mothers and babies and helped make baby clothes.

To keep everyone healthy, families also relied on herbal medicines and on local healers who used an array of folk remedies to varying degrees of success. Over the years, I have written down many of these treatments. Mr. Arlie told me, "We used catnip tea for a teethin' baby. Blackberry juice was used for an unsettled stomach. The inner bark from yellow poplar was steeped in whiskey and used as a remedy for yellar janders [jaundice]. For coughs, the inner bark from wild cherry or black cherry was used to make a cough syrup. They made a pine resin and hog grease poultice that was used for bad backs. If you had an infected area, they took the peach leaves and put them in a cloth and beat 'em up and put that on the infected area."

Other traditions in this farming community revolved around the hospitality of welcoming visitors and travelers who made their way across rural back roads, even though residents were skeptical of strangers. Mr. Homer talked about how difficult travel was for people up until roads were paved in this area in the 1930s. There were no restaurants or hotels, so travelers had to rely on the generosity of residents along their route for a safe place to stay. Passing travelers shared information with each other along the road about where springs and watering holes could be found to water their horses and teams.

During the last century and well into the 1900s, travel between the towns of Franklinton and the Bogalusa area, thirty miles to the east, or Mandeville, sixty miles south, was a major journey by today's standards. Roads, when there were roads, were often filled with obstacles such as downed trees or soft, muddy, wagon-wheel ruts that were difficult to pull wagons over. Before the days of modern stores, obtaining even something as simple as thread was a challenge. One woman recalled her mother leaving money and a request for thread in the mailbox, which was obligingly filled by the mail carrier. A number of peddlers also traveled these routes, selling their wares to families along the way. Mr. Seldon recalled that these travelers would stop at a place known as "Peddlers Spring," which was located halfway between Franklinton and Bogalusa along what is now Louisiana Highway 10. There, fresh spring water flowed from a sandy, red-clay bank on the north side of the road. This water provided a welcome refreshment to peddlers, who would stop there and set up a makeshift market.

Peddlers were not the only ones who enjoyed the cool spring waters. Other travelers would pause to rest at the spring, and to pass the time or to exchange news with acquaintances they rarely saw. In those early days before the telephone or other modern means of communication, common meeting places such as this offered some of the few chances to get news from people in the community or the outside world. Over time, travel became easier and merchants opened stores in towns, and the peddlers disappeared. Eventually, the spring ran dry.

Another landmark on the route between Franklinton and Bogalusa was a place called "the blue water hole." From his childhood, Mr. Seldon recalled his father stopping their wagon to water their horses at the popular spot when they were making trips to town. The blue water hole had a well-known resident, a gigantic brown-blotched water snake, or gopher snake, as some call it. The serpent was a familiar sight along the water's edge, and many considered it the guardian of the watering hole. "When you saw it by the water," Mr. Seldon recalled, "somehow you had the feeling that all was safe to water horses and livestock." He reminded me that that watering holes like these were as important as a gas station is along a modern-day highway.

Before roads were paved in the 1930s and 1940s, travel could also be dangerous due to unsavory characters who could be found along the way, especially during the particularly lawless period toward the end of the nineteenth century. Many old-timers told me stories of some of the more notorious outlaws in the region, particularly the infamous Eugene Bunch gang. Between 1887 to 1892, Bunch and his men were rumored to have stolen several hundred thousand dollars from trains across the South. Bunch's home in the swamps of the Pearl River in rural Washington Parish proved to be a good hiding place from the law, for a while. Eventually, Bunch's fortunes ran out and the train robber was cornered near his home and shot dead in a hail of gunfire.

Whether it was the lack of lodging or restaurants or the fear of human predators, people traveling on foot or horseback often found themselves knocking on the doors of farms and homes along the way. Although many residents in the area were understandably skeptical of outsiders, most would welcome a visitor, saying, "Go ahead an' light ah hitch an' come set a spell," meaning tie up your horse and come visit on the front porch.

Peas on Washington
Parish sidelin red
clay sidelin ground
ole Pecan tree,
I see'd one of em was
holler.

22 Oct —
"Them's put up Peas,
you can't buy em"
Anybody can't plow
ole Beck on these here
vittles hoss, they'z
somethin wrong with em,

This is one of the many sketches I made in 2007 while traveling the backroads of the parish.

Many of my favorite times in Washington Parish consist of sitting on a porch or swing and just looking out on the beauty of the land, while listening to stories told by my friends. Alongside this undated sketch, I wrote, "It is at times such as this that oral histories and legends are passed on to me for preservation."

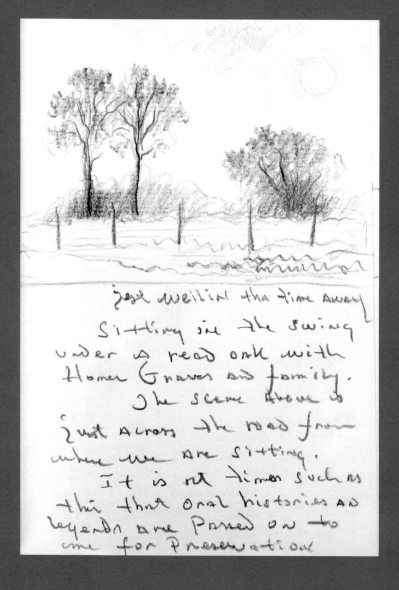

Just Waitin' the Time Away

Sitting in the swing under a red oak with Homer Graves and family.

The scene above is just across the road from where we are sitting.

It is at times such as this that Oral histories and Legends are Passed on to me for Preservation

In the early part of the twentieth century, the US government made an effort to count the number of people living in the nation's most rural areas. From time to time, census takers would make their way through Washington Parish's back roads to interview wary farmers. A great old story relayed to me by Mr. Robert Toney recounts his neighbor's interaction with a census taker in the 1920s. As the story goes, his neighbor assumed the census man was from New Orleans, which was likely the only city he was familiar with.

"What you say you was ah taken'?" the neighbor inquired.

"I'm taking the census," the stranger replied.

"I ain't got much senses, I need what I got," the neighbor shot back.

"Not those senses," the somewhat embarrassed stranger answered. "I need to know how many people live here. How many children do you have?"

"Well, I don't rightly recollect right off. You can come on in en count em if ya' will. They's a room full of 'em thar in tha cabin."

Being the hospitable man that he was, the neighbor insisted that the census taker "draw up a chair on the porch and set a spell." In addition to the livelihood he derived from farming, the neighbor was also known to practice the art of "running moonshine." His still was a well-kept secret cleverly hidden deep in a swampy area some distance from his cabin. The neighbor called to his oldest son, who was out chopping up stove wood, "Boy, go out to the shuck crib and fetch that gray ceramic jug." The boy did as he was told, brought the jug, and with a dull echoing "pop" the cork was pulled from the jug. The first swig was taken by the neighbor. Taking the first drink was often customary among moonshine distillers. It was a gesture showing that their whiskey was pure from poison, or "pizen" as the storyteller called it, and that the distiller was proud of his product. "Now there, Mr. Census Taker, take a draw off this jug. This here is as fine ah liquor as you ever tasted," the neighbor boasted. The census man politely refused at first, saying he was unaccustomed to strong drink, but appreciated the offer just the same.

After a few more minutes of talking to his host, the visitor decided that one sip couldn't possibly hurt anything. He raised the jug to his lips and took a healthy swig. "This stuff really burns," the census man remarked. "But how warm and pleasing it is going down. Say, I don't mind if I do have another sip." After several more sips and conversation, the census taker asked the neighbor if he might not have some of these fine spirits for sale. Again, the son was called from his work to go and fetch another jug filled with "white mule." The boy handed the jug up to the census taker, and father and son watched from the front porch as the New Orleans stranger closed the picket-fence gate and started up the dusty road, weaving his way in a less than straight direction, his book in one hand and a jug of white mule in the other. The neighbor predicted the census taker would soon come calling again for a refill.

Moonshining was a popular business in this region. There was also a lot of crime associated with moonshining, including the murder of two sheriff's deputies, Wiley Pierce and Wesley Crain, on March 3, 1923, in Betsy's Creek Swamp, a remote area south of the Sheridan community. The deputies had been investigating an illicit moonshine-distilling operation. Their bodies were found tramped into the mud near the creek. Eventually two men were charged and convicted in the killings. John Murphy was executed by public hanging in Franklinton. Fact or legend has it that there were no less than seven

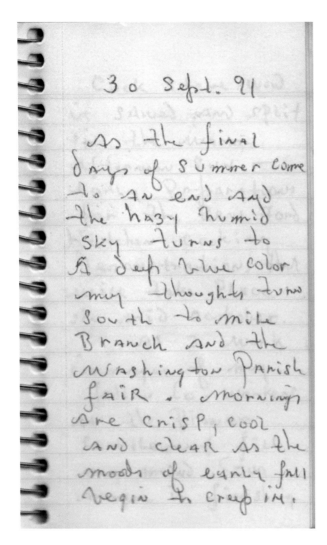

A journal entry from September 30, 1991, looking forward to the Washington Parish Fair.

stills running in full operation at the time deputies Crane and Pierce made their raid on the Betsy's Creek still. Some claim the stills were running with the full knowledge of higher-up officials who had certain financial arrangements with the distillers. Before his hanging in 1923, John Murphy's last words were, "If I told everything that I know there would be a lot more widows and orphans in Washington Parish." A second man, Gideon Rester, was given a life term at the state penitentiary, where he was killed in a prison fight. Washington Parish was an ideal place for moonshining, with the parish's remoteness, its abundance of corn and sugar cane for the production, and its desperately poor residents. One of my friends told me, "Poverty drove 'em to it. Starvin' little mouths to feed."

Some early visitors to Washington Parish ended up staying, including a few immigrants from Italy and Eastern Europe. One well-known settler in Washington Parish was a man named George Kostov, from Bulgaria. Just how he wound up in this region is unclear, but he arrived around 1910 to begin his new life as a homesteader outside of the settlement of Sheridan. He came with little more than the clothes on his back and a few prized possessions from his homeland. Importantly, he brought heirloom seeds passed down through his family for generations. The seeds included paprika, several types of tomatoes, and Bulgarian hot peppers, one in particular a large variety of red pepper. To avoid detection and possible confiscation, George had the seeds sewn into the seam of his coat and hidden in his hatband. These seeds served as a living reminder of the land and heritage he left behind. They symbolized a link with his past and a new beginning. Like the seeds he carried with him, George was also being transplanted.

After settling in Washington Parish, George married and began farming. He also grew a small amount of tobacco for his own use, which he cured with a sprinkling of syrup water. George smoked this home-cured mixture in an old curved pipe that he referred to as a "beiler." He often walked the four miles out to the old general store and post office at Sheridan to get his mail. Many old-timers recall that you could see smoke from George's pipe long before he came into view on the road.

Most often, visitors were extended-family members, friends, and neighbors who would spend an afternoon and enjoy a meal together after their journey. Late-afternoon conversation turned into an invitation to stay the night, as it did to one of my favorite families, Preacher Blount and his wife. "No Louisiana visit is complete without a visit to Preacher Blount's over in Springhill, Louisiana, to buy ribbon cane syrup," I recalled in a 1989 journal entry. "A quick stop to buy syrup turns into a several-hour visit. He even has Mrs. Blount make biscuits at 10:30, 'Stay the night, we've got plenty of room.'" Some visitors, inevitably, wore out their welcome. There was one gentleman who lived in the community who was known for showing up at neighbors' homes around mealtime. Always welcomed at the table, Mr. Jim Bolware accepted invitations to stay by saying, "Well, I don't aim t' disfurnish ya' none, but I believe I will."

In all of my many visits to the Washington Parish homes of friends, neighbors, and even strangers, without exception, I have always experienced such a welcoming level of hospitality. It is a community that has taught me that we are all strangers in this life in some sense, and that the gift of conversation and generosity of spirit is the best we have to give. I feel lucky to have been the recipient of such a gift.

FALL

I am a creature of habit, and as summer turns to fall each year, I start looking for my suitcase. August is "Garden Month" at Blackberry Farm. This is my busiest season, but when I feel a change in temperature, my fall travel plans take precedence. For more than forty years, I have set my entire calendar year around my annual autumn pilgrimage from my home in the Smoky Mountains of Tennessee to Washington Parish, Louisiana. I time my trip to coincide with the Washington Parish Fair, which happens during the third week of every October.

In recent years, I have switched things up slightly to begin my fall season with a trip abroad to visit friends I've kept up with since I was a teenager. In 1961, when I was eighteen, I was invited by a family friend and faculty member of the University of Tennessee Department of Music on a tour of England, Scotland, Holland, France, Germany, and Austria. On that first trip to Europe, I met and befriended Oskar Korber while staying in Seeham, Austria. We were born a month apart, and like me he was a painter, but he was also a musician. Over the years, Oskar opened a door for me to a world with fascinating people, and we remain the closest of friends.

In 1970 after undergraduate school, I returned to the region around Burgenland Province, south of Vienna. I lived there two years painting and drawing and working in vineyards, and as a farmhand. I was enchanted by the area's agrarian lifestyle, and even today I remain captivated by this bucolic region, which is remarkably unchanged by modern farm methods. Since 2010, each fall I travel back to visit Romania, Hungary, and Austria to attend the annual Tschardakenfest Corn Crib Festival in Halbturn, Austria. I always visit my friends there, old and new. Just as I have written journals documenting life in Washington Parish, I have also documented parallel lives in Austria and Eastern Europe, in dozens of additional volumes of journals. On those pages, I often find myself reflecting on how similar everyday living is in that rural corner of the world to the lives older generations have told me about for decades in Tennessee and in Washington Parish. I think about the vast distance and two distinctly different cultures, and yet I am mostly drawn to the similarities and commons bonds that give us kinship.

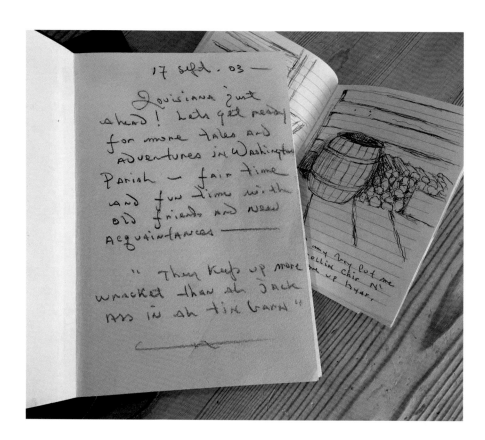

The handwritten journal entry reads:

17 Sept. 03 —

Louisiana just
ahead! Let's get ready
for more tales and
adventures in Washington
Parish — fair time
and fun time with
old friends and new
acquaintances ——————

" They keep up more
wracked than ah Jack
ass in ah tin barn "

Every year, I look forward to
my trip to Louisiana with great
anticipation.

The minute I set foot back on American soil, I am off on the next leg of my October journey. I have barely enough time to wash my clothes, grocery shop, and repack before I hit the road for Louisiana and the Washington Parish Fair. The rituals of my early visits still play in my mind, just as they did in this October 1991 journal entry: "3 a.m. departure—The air is still damp and cold. All is quiet and hushed as my journey into black night begins, just me and my thoughts of the journey and people that lie ahead. As I travel on into the night it almost feels as though I am flying a plane alone in this small cockpit flying towards my destination far to the south. The only light is that of my headlights illuminating the long road ahead. At times the darkness becomes a screen onto which images of Washington Parish are projected. It's early afternoon as I turn off Highway 10 and head up Mr. Seldon Lang's gravel road. As I pull up to the house, Mr. Seldon is sitting on the old swing waiting for my expected arrival dressed in faded blue overalls and a white T-shirt. There is the usual exchange of news, talk of the long journey and the passing of another year. I can even hear a soft wind whisper through the pine needles as our reunion continues. Soon Mrs. Lang calls from the kitchen, signaling that afternoon coffee is ready. This is no ordinary coffee. It is strong Louisiana coffee, which quickly jolts any remaining drowsiness from the long day's journey. Along with the coffee there is always homemade cake of some sort to go with it."

An October 23, 2009, journal entry:

For the most part I have been on the road for 23 days. On the first of October I left Tennessee for Romania where I stayed for 15 days, and now continue here in Washington Parish, Louisiana. The drawing detail of the logs to the left closely resemble those that I saw on Romanian log houses in Maramures.

Here at the Mile Branch we are celebrating the history and heritage of this area at a collection of old homes, school, gristmill, barns, store and other structures. These relics represent a piece of our long history since departed. Five days ago, I was living in the midst of settings similar in many ways to Mile Branch with one large difference. Here it is a museum piece to be visited and celebrated once a year. There it is history, heritage, and a way of life being lived here and now. Four days ago I was walking down country roads in Romania, experiencing a way of life that has long since disappeared from American memory. They're the old ways and everyday normal life, unlike here when some old-timer says, "I remember how we did it back in the old days." There today they are still living the old days. It is not some distant memory but a way of life that is still being practiced. I was most fortunate to spend a number of days in the midst of that way of life. I hope to experience it again in the not too distant future.

Returning this year to the Mile Branch Settlement my first observation was how similar some of the old buildings always are. The picket fences were the same as many that I saw there and many of the old farming methods are the same. The similarities in the mountains of Tennessee are more evident as in the countryside in mountain terrain. There is a natural kinship between the two places that I was instantly aware of and drawn to. Our ways of life are quite different and yet they are closely related. When I go to bed at night I can still hear the voices of those people that I spent time with.

23 oct 2009

for the most part, I have been on the road for 23 days, on the first of october I left TN for Romania where I stayed for 15 days, and now the journey continues here in Washington Parish Louisiana. The drawing detail of the logs to the left closely resemble those that I saw in Romanian log houses in Maramures. Here at the mile Branch we are celebrating the history and heritage of this area at a collection of old homes, school, grist mill, barns, store and other structures. These relics represent a piece of

For more than forty years, I have worked as a volunteer at the fair. I have spent countless hours on the porches of the old farm buildings, recording stories I hear from the people I meet. *Photo from author's collection.*

At the heart of the Washington Parish fairgrounds is the Mile Branch Settlement, where pioneer-era buildings from all over the parish have been clustered to form a little historical village. During fair time, Mile Branch is run by pioneer reenactors, community volunteers, and close friends and neighbors who have been staffing it for generations. For more than twenty-five years, my job has been cooking on a woodburning stove in the Ben's Ford Kitchen.

I knew the first time I visited the Ben's Ford Kitchen that I just had to be a part of this fair. I remember the smell of fried apples, sliced fried sausage, and coffee cooking on the woodburning stove. I introduced myself to longtime volunteer Mrs. Eileen Crane. She shared the

I think of my fall travels to Louisiana as central to my life. In an entry from October 2014, I describe it this way:

Fair time in Washington Parish for many years now and nothing has ever come close to matching my social calendar in importance than the third Wednesday of every October. Don't get married on that date, I won't be there! In fact, don't plan anything and expect me to be there. There is only one place I will be in and that is Franklinton, Louisiana, for the opening day of the Washington Parish Free Fair.

I fell in love with the Washington Parish Fair that first time I attended in October of 1988. All over the United States, communities hold country fairs, even in Tennessee, but I find the Washington Parish Fair to be unique. I've tried to explain the difference to friends of mine. To me, a trip to this fair is like stepping back in time, where the traditions of an agrarian past are very much alive.

Mile Branch Settlement.

Volunteers such as Bevie Simmons (*left*) and Eileen Crane (*right*) staff the historic homes at the fair's pioneer village. Mrs. Crane was instrumental in my becoming a volunteer at the Ben's Ford Kitchen in 1988. *Courtesy of the Varnado Museum.*

history of the kitchen building with me. The log structure was built in 1855 in the Ben's Ford Community close to the nearby town of Bogalusa and had been moved to the Mile Branch Settlement in 1983. It had all the original furnishings, including an old chicken coop affixed to the exterior wall. She invited me to be a volunteer the next year, and ever since, I have cooked three meals a day on that wood stove, using recipes I learned from local cooks, all reminiscent of what farm families in past generations would have eaten. Over the years, I have made whatever comes to mind with whatever I find fresh at Jack Brown's Supermarket in Franklinton. I once told someone that I intend to continue working at the fair as long as I am able. Then when I am no longer able, I will still continue to work there anyway.

While the fair is the centerpiece of my Octobers, the entire ritual of my annual journey is like a favorite movie that plays on a screen again and again. One of the best parts of this pilgrimage to Louisiana is the drive down. A lot of people don't like driving, but I love every single mile of the way. When I was a younger man, I would

I have always been a list maker. In my preparations for my annual trip, I always make grocery lists and to-do lists so I don't forget anything. Here is a September 20, 1990, journal entry:

I called about the fair, it starts October 17 and ends on Saturday, October 20. I need to be down on Tuesday the 16th to set up the kitchen with decorations. We will meet at the fairgrounds at 8:30 a.m. A list of things that I will be taking:

1. Canned goods
   Field peas
   Cornfield pole beans
   Tomatoes
   Kraut
   Hot peppers
   Beets
2. Dried herbs, sage rosemary thyme tied in bunches to hang on the wall
3. Braided garlic strand, cayenne peppers, dried white corn tied to hang on the walls
4. Louisiana sweet potato pumpkins
5. Kitchen large and small cast-iron pots spoons ladles any useful old item iron skillet
6. Long narrow woodbox for stove wood
7. A small bundle of lighter wood as part of the display
8. Dress for the fair, same as last year, blue denim shirt and overalls
9. Menu for woodstove cooking demonstrations
   red beans and rice
   field peas and rice, smothered cabbage
   small butter beans
   white butter beans with tomatoes
   chicken stew, brown roux, onions, garlic, bell pepper and serve with rice
   chicken stock with okra and tomatoes
   gumbo
   iron skillet fried potatoes half runners with new potatoes cooked on top
   any ole-time recipe that comes to mind

Take spices along, bay leaves, thyme, pepper, cooking oil, streaked meat, cooking items needed on Tuesday the 16th.

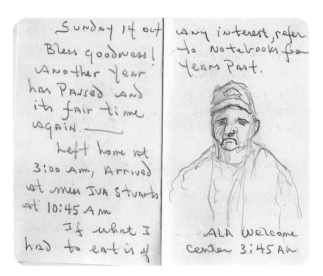

Sunday 14 Oct
Bless goodness!
Another year
has Passed and
its fair time
again.———
Left home at
3:00 Am, Arrived
at miss Iva Stuarts
at 10:45 Am
If what I
had to eat is of

any interest, refer
to Notebooks from
Years Past.

ALA Welcome
Center 3:45 Am

A 2013 journal entry referencing the meal I eat every time I visit Mrs. Iva Stuart's restaurant.

leave home around three o'clock in the morning. I have favorite routes along meandering country roads that I could travel blind. I look forward to the sun coming up as I hit the Alabama Welcome Center just short of the Mississippi state line. Then, as the early morning fog lifts from southern Mississippi farms and stretches of pine timber, I finally make my way through the back roads of Washington Parish.

Once in Louisiana, I always stop for lunch in the tiny town of Angie, at Mrs. Iva Stuart's Café. I order the same meal every time and it never varies. It's always fried chicken, rice and gravy, potato salad, turnip greens, cornbread, and lemonade. The only change to this lunch is the possible addition of a slice of sweet potato pie, if Mrs. Iva has any left. Every single highway, every back road, every mile, and every meal is a cherished part of my trip. Of course, as I get older, it has grown hard to make the trek all in one stretch. Now I break it up with an overnight stop, but the milestones are the same. Fi-

nally, when I cross the bridge into Franklinton over the Bogue Chitto River, I know I am home.

Once I've reached Washington Parish, my first stop is always a visit with my good friends, where I am never without my pencil and notebook. For decades, I stayed at the Vise family home, and it is through Mr. Calvin Vise that I have met so many of the people I have come to regard as extended family members. Every time I'd arrive, there was all this wonderful food waiting for me. Mrs. Rose Vise was one of the greatest cooks I have ever had the pleasure of meeting, and she'd have a chicken pie or huge pot of gumbo on the stove. With nine children, she didn't know how to make a small portion.

From there, I would visit Seldon Lang or Homer Graves, who have both now passed on, and Arlie O'Bryant. I spent long hours recording their stories, often sitting in silence as they recounted the details of their lives and history lessons from the parish. I never wanted to miss an opportunity to talk to them. I felt then, as I do now, that they were the masters and I was the student. For decades, I worked alongside these men planting fall gardens and observing and learning from their extensive farming knowledge. Before heading out to the fair, if I was lucky, Mr. Homer would take me on long drives through the backwoods parish roads, pointing out landmarks and sharing stories of local families and their farms.

Fair week starts on Monday with a big banquet for the volunteers. The event is put on by the Washington Parish Fair Association. By Tuesday morning, I am at the fairgrounds among the pine, sycamore, bay, and oak trees and dropping off supplies, chopping firewood, and getting ready for opening day on Wednesday. The fair begins with a parade that winds through town and ends at the gates of the fairgrounds. The parade features marching bands and cheerleaders from areas schools,

Catching up with my friend Calvin Vise.

local 4-H clubs, cowboy clubs, vintage tractors, and much more. You name it!

Once the fair starts, it is chock-full of community traditions. Young and old join in the art, writing, livestock, agriculture, and floral competitions. On numerous stages, there are live performances from musicians from all over the region. I love the old-time fiddle players best. While he was living, Mr. Seldon Lang and I would walk down to the stage area at around 10 o'clock every Saturday morning of the fair, where we would find ourselves good seats on the front row, so we could enjoy the old fiddle melodies up close and personal. Mr. Seldon was a wonderful fiddle player, like his father before him. My friend used to perform in the fiddler's contest that was held every fair Saturday. Even today, the fiddlers play all of the old-time favorites, "Maiden's Prayer," "Faded Love," "Over the Waves," "Kentucky Waltz," and "Soldier's

Joy," among others. Often the musicians perform in the Half Moon Bluff Baptist Church located in the Mile Branch Settlement, where rousing choruses of favorite hymns resonate through the crisp fall fair days.

The large midway features thrill rides and games, as it has since the fair's earliest days. In the evenings, there are youth and professional rodeo events, fair and rodeo queen pageants, and concerts. During the day, you'll find the "beautiful baby" competition. For the Senior Ladies Promenade, older women in the community wear hand-sewn pioneer dresses, and are escorted across a stage, pageant-style. They are interviewed about life in an earlier era, and some of their responses have been extremely entertaining. One elderly woman was asked if she'd ever been bedridden. Her answer amused the enthusiastic crowd. "Hundreds of times!" she exclaimed, "and twice in a wagon."

One of the most enjoyable parts of the fair is the music, especially the old-time fiddle music that is so popular in this piney woods region. I love to hear those traditional tunes and have enjoyed listening to them played on the porches of the Mile Branch.

These photos show the fair in the 1920s and today.

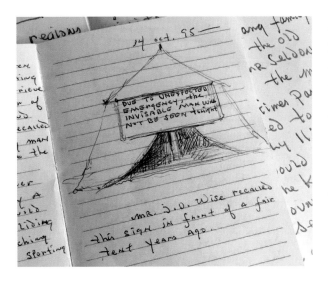

An October 14, 1995, sketch and journal entry of a fairground tent set up in the early days of the fair.

Historically the fair had a number of attractions like high-wire acts, magicians, jugglers, hypnotists, and contests designed to showcase locals' strength and skills. Old-timers talked about the harnessing contest, where the entrant who could harness mules and horses the fastest would win a pair of boots or barrel of flour. There was greased-pig catching, hog-calling, and hog-tying. In the early 1900s, when oats and other grains were still being cut with cradle scythes, Mr. Seldon Lang told me his father, Mr. Albert Lang, was considered the best at cradling grain, and he would demonstrate his skills for audiences at the fair.

The whip-popping contest measured how well a contestant could handle a whip. In the early 1920s, a man named Calvin McNeese was supposedly the best of his time. McNeese could crack a whip louder and harder than any challenger. His whip would hit the ground with such lightning speed, it could set the grass on fire. There were wood-chopping contests, and even a greased-pole-climbing contest, where a five-dollar gold piece was placed at the top of the pole. Young men took their turn scaling the pole to win the prize. There were near successes and humorous failures as more than one young man lost his grip and plummeted to the dusty ground below.

Many of my favorite stories that I've recorded have come from the fair. Each day, I get to the fairgrounds early in the morning as the sun is coming up. I light the woodstove and start cooking breakfast, usually baking biscuits and frying what the old-timers called "three-tiered bacon," or slab bacon, meaning three streaks of meat between fat. Even though I am not a coffee drinker, no breakfast here would be complete without strong black coffee, like the kind Mr. Seldon used to drink. He always told me that "it doesn't take nearly as much water to make coffee as most people think."

Morning coffee is the best way to catch up with friends you haven't seen since the last fair. My first conversation of the day is usually with Gus Magee, the craftsman who hand carves coveted Tupelo-wood dough bowls. He also weaves large white oak split baskets, among many other things. Gus has a booth in the Mile Branch Settlement where he sells his woodworking crafts and children's wooden toys. Over the years I have collected many of his creations, and they are some of my most treasured possessions. Gus was also the source of the elusive Unknown Pea of Washington Parish that, for decades, I had searched for. He is a wealth of stories about Washington Parish and Magee family history.

Another regular who would drop in during the early hours before the fair opened was Mrs. Letha Toney. Up until her passing at age ninety-two, she would build fires in all of the cabins and light the kitchen stoves around Mile Branch Settlement. I always felt that Mrs. Letha walked right out of an earlier time, when these old

My good friend Gus Magee is well known for his woodworking talents, making everything from dug-out canoes and dough bowls to woven baskets, children's toys, birdhouses, and rocking chairs. Over the years, he has given me a number of beautiful split-wood baskets and wooden bowls. I use them every day and think of his artistry and our friendship. *Photo from author's collection.*

Before her passing, Mrs. Letha Toney was one of my best friends at the fair. Over the years of our volunteering together, she shared with me many stories of her childhood, and she was one of the best resources in tracking down the elusive Unknown Pea. *Courtesy of the Varnado Museum.*

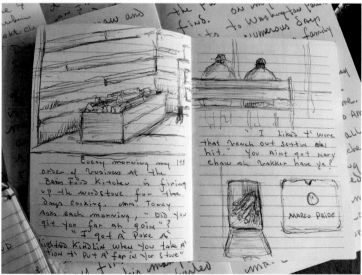

Porch and kitchen scenes from the Ben's Ford Kitchen, October 1995.

buildings were still in full-time use. She was full of stories about her childhood, when her father farmed over one hundred acres of cotton, sugarcane, sweet potatoes, and the infamous Unknown Pea of Washington Parish, which she first described to me. Even as a young girl, Mrs. Letha would tell me, she was taught every job on the farm and worked at every aspect of farm life.

My next order of business is starting dinner. My pioneer days–inspired menus and recipes often include baked sweet potatoes, smothered cabbage or greens of some sort, and a local variety of field pea or butter beans cooked down with smoked meat. Sometimes I make chicken stew, beans and rice, or skillet-fried potatoes.

CHICKEN AND OKRA STEW

Makes 10 servings.

—◆—

I once tried to call this stew "gumbo" but quickly learned that Louisianans are extremely particular about what constitutes that regional delicacy. Rather than serve an imposter, I realized that this tasty and hearty soup is just that, tasty and hearty enough to take the chill off. I like to serve it over rice.

1 tablespoon bacon grease or extra virgin olive oil

2 small yellow onions, chopped

2 garlic cloves, chopped

1 pound dark-meat chicken pieces, with bones

1 (14.5-ounce) jar or can of tomatoes, with juice

½ pound fresh okra, sliced

1 small dried red pepper pod, crushed

1 bay leaf

2 quarts or so of chicken stock or water

Hot cooked rice for serving

1. In a Dutch oven set over a medium-high fire, add bacon grease and sauté onions and garlic until soft and golden, about 8 minutes.
2. Add chicken, canned tomatoes and juice, okra, crushed pepper, and bay leaf. Pour in enough stock to cover the ingredients.
3. Simmer until chicken is cooked, about 1 hour. Add stock as necessary. Remove chicken pieces and debone them; then return to the stew. Remove bay leaf. Serve hot in a bowl over rice.

SMOTHERED CABBAGE

Makes 6 servings.

—◆—

As simple as it is, there is nothing more delicious than smothered cabbage. It is an easy and inexpensive side dish that is always a hit among fairgoers.

1 head of cabbage

1 tablespoon bacon grease or extra virgin olive oil

4 cloves garlic, chopped

1 fresh hot red pepper, sliced, or ½ teaspoon
    red pepper flakes

Ground black pepper, cayenne pepper, and
    salt to taste

½ cup water

1. Chop cabbage into large pieces and set aside. Heat a large saucepan or Dutch oven over medium heat and add bacon grease. Sauté garlic until tender, about 2 minutes.
2. Add fresh red pepper, black pepper, cayenne, and salt. Add the cabbage and sauté until it starts to get tender. Add water, put the cover on the pot, and allow the water to steam the cabbage until soft, about 15 minutes. Serve warm.

SKILLET FRIED POTATOES

Makes 6 servings.

———◇———

There are a million ways to cook potatoes, and I have had many people share recipes with me. One of my favorites came from a woman I met at the fair named Nita Duncan, who sliced and dredged hers in flour and fried the slices in butter. Here is another favorite way to fry potatoes, and this is how I often cook them at the fair.

4 to 6 large red, white, or yellow fleshed potatoes
(not Russet or Idaho baking types)
Vegetable oil (enough to come up a half-inch in your skillet)
Louisiana seasoning salt

1. Cut the unpeeled potatoes into ¼-inch slices. Heat oil in a large cast-iron skillet until it is about to smoke, and add potatoes. Turn potatoes occasionally until golden brown on each side.
2. Remove the potatoes from the oil and drain on paper towels. Sprinkle with seasoning salt and serve hot.

Once dinner is served at midday, I am ready to make my standard rounds in the Mile Branch Settlement. I always visit the cracklin' pen where volunteers render lard and fry up those hot, delicious pig-skin delicacies with just the right amount of seasoning. I visit the country store, which sells soap, sliced hoop cheese, and baked goods. Then there's the cane grinder, sugarcane evaporator, and the gristmill for cornmeal grinding, which is always manned by Dale Graves, Mr. Homer Graves's son.

Late afternoons and early evenings at the fair are also the perfect time to wind down, sit back, and bend my ear for tall tales, good old jokes, and family stories of life in the parish. This is also the best way to do detective work

Journal entry with a recipe for my friend Nita Duncan's fried potatoes.

when looking for seeds that families might have stored at home. I'll ask people I meet to bring me samples of seed, or I make contacts and follow up with them when I get back home. Most often, I meet fairgoers who are just passing through the Ben's Ford Kitchen. They might be stopping for something to eat and will reflect on their own family's history, inspired by the setting. I do my best to capture the details they share in my journals. Sometimes that chance encounter is the one and only time I'll ever have to talk to these characters. I barely have time to jot a few things down or sketch their faces, and they're out the door before I can ask them more. Other times, they become dear friends that I expect to see each year.

I remember one night, a fellow volunteer, Miss Bevy Simmons, and I were closing up and this little old lady who was thin as a rail shuffled in. She started talking about growing up in the Honey Island Swamp nearby, just her father and three sisters. The family lived in a board-and-batten house on pilings in the swamp. Their father trapped gators and all kinds of things to support them. She said, "Paw didn't allow no boys around when us girls wuz growing up. We wuz nigh grow'd before we knew what a boy wuz."

I often hear ghost stories from the area, like the one about the haunted Child's Bridge that crossed Pushapatapa Creek. Mr. Arlie O'Bryant told me about his brother's terrifying experience crossing the bridge late one night in the summer of 1934. Berkeley O'Bryant was riding his horse along the Child's Bridge Road on a dark moonless night. As he approached the bridge, he was seized by an eerie sense of fear. With no warning, just as he started to cross the bridge, he felt something light on the horse's back, and he could hear his shirt being ripped. He felt the burning pain of something slashing away at his back and neck. The terrified horse reared up and almost threw Berkeley out of the saddle and

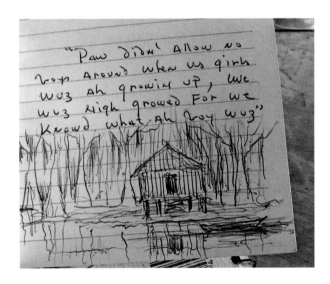

October 23, 1994, journal entry and sketch detailing the story of a woman who grew up with her father and sisters in a remote nearby swamp.

over the bridge, into the creek below. Berkeley took off at a fast gallop toward home. Upon entering his house, Berkeley was horrified to discover that his bloody shirt had been torn to shreds and his back disfigured with deep cuts. Berkeley's father listened to his son's accounting of the ordeal in silence and pondered the details of the attack, after which he plainly remarked, "Well, it could've been so."

A few good stories from my fair days revolve around well-known characters that once lived in this region. One of the most colorful of those is "Fighting Pierce Parker," and Mr. Seldon had numerous stories about him. Pierce Parker had been a Confederate soldier and lived in Washington Parish following the Civil War. He was well-known as a champion in what was called "fist and skull knuckle fighting," which is just what it sounds like, bare-knuckle fist fighting. Parker was a fierce fighter, savagely bloodying opponents who came from near and far. One notorious match ended with his vanquished challenger yelling "Butterspoon!" or the equivalent of "Mercy!" to stop the carnage. Parker's fighting ways made him many enemies. When he was up in his eighties and in poor health, one of those enemies passed him on the road and jumped off his horse to attack him. Parker mustered the strength to fight back and brutally beat the man to within an inch of his life. Eventually, Parker died in prison in his eighties after being convicted of shooting and killing a man he felt had insulted him.

Another of the legendary characters from the parish was a woman named Caroline Gabe, who lived alone and farmed a sizable piece of land. Homer would tell me, "Aunt Caroline sure could plow a mule, Hoss! She'd plow it 'til it fell. She raised about six acres of cotton and two acres of corn. Come cotton pickin' time, she picked her own cotton and she didn't want no help. Come time to pull corn, she'd take them two mules and

away at home and it is a long time until next years fair. I have had many such dreams where all of the scenes seem to blend into a single experience.

Sweet Potatoes in the warming clauset of the Marco Pride woodburning stove

19 oct
I cranked up an we ground another 100 lbs, we're gona be behind shor-nuff tomorrow.

LA and MS heritage seeds —
SNOW ON THE Mt. BUTTERBEAN 1880
PURPLE HULL WHIPPOORWILL
WALTHALL CO MS 1860

Journal entries depicting Washington Parish Fair activities in days past.

we were still picking cotton when fair time came around. When the fair started we were finished with biggest part of the picking.

It was always tradition for everyone to have a new outfit for the fair.

A 2001 sketch and journal entry about picking cotton.

wagon down to the cornfield. She'd fill them white oak split baskets and when they was full, she'd throw them into the wagon and haul that corn off to the barn. She didn't want no help." Homer Graves said he once asked her if she was afraid of living alone. She replied, "Not for a minute. I've got this old hog leg [a sawed-off shotgun] under my apron, and I ain't afraid to use it." Another fiercely independent woman, Ana Childs, lived deep in the swamp. She was best known as a crack shot and an expert hunter. She trapped all manner of creatures and sold or traded the skins to support herself. She was also well-known for making a special variation of pecan pie, substituting hickory nuts for pecans. Mr. Homer Graves recalled having once sampled the widow's hickory-nut pie and described it as a delightfully unique dessert.

Many of my older friends recalled the days when their families were still harvesting crops well into October. Once the corn and particularly the cotton were ready to pick in the fall, everyone pitched in. Many old-timers would often talk about their experiences picking cotton as young children, having smaller bags specially fashioned for them to handle the heavy loads. "We didn't have a cotton picker back then." Mr. Homer called the crops "people picked," and some of my friends' earliest memories involve picking cotton, sometimes around the clock, especially when the light of a full moon helped illuminate the white cotton bolls. As Mr. Homer told me, "You'd try an git in that cotton patch es early as you could, an pick cotton while it was damp so hit'd weigh better." Cotton wet with dew always weighed more at the mill, fetching a higher price. "We finished up just about daylight and we picked a bale a cotton and got it to the mill," Mr. Homer remembered. "We picked right at three hundred pounds, and after we got done with picking that morning, we went back out and picked the scraps that was left behind. We used the cotton scraps to make us our first mattress." Cotton scraps were used for everything from stuffing pillows, to spinning yarn, or making batting for quilts. Nothing went to waste. Parents often promised a trip to the fair as a reward for a long night of cotton picking.

Families pulled together to get cotton picked. Hired workers were paid a meager wage to gather what children and extended kin could not. Many recalled seeing large fields of hot cotton pickers making their way down vast rows, picking and filling long sacks that dragged behind them. When the workers filled a sack, they would pull it to a cotton scale that hung from a large wooden tripod. Workers were paid per pound picked. Once weighed, the sacks were emptied into large wagons with tall sideboards, and they were then driven by mule to the gin. Some of my friends had childhood memories of riding to the cotton gin, half-buried in the cotton on a cold frosty morning, and waiting in a long line for their cotton to be off-loaded and sold.

Many talked about how fall was the only time of year when their families ever had any money to spend, because it was the time when they would gin their cotton, take their corn to the mill, and make cane syrup and sell it. "That was when we paid off our bills and we had some leftover to purchase a few essentials that we couldn't make or raise on the farm," Mr. Seldon recalled. "Sometimes we made a little extra if Paw sold a yearling, some syrup, or a few hides that we trapped out in the woods and swamps."

The fact that the Washington Parish Fair was, and still is, a "free" fair, meaning that it does not charge admission, allowed families with no spare money to take part in this important community event. In the old days, most families brought their own picnic meals to eat

An October 1987 journal entry:

I arrived for a visit with Paw Lang just in time to sow oats. He was disking up his deer patch down in the woods. With dark clouds moving in we finished just ahead of the rain, the first in two months. The fields were like a dustbowl. Paw said this weather and drought was the worst he had ever seen. The open plowed field, pines in the background, and a threat-ening sky with Paw sowing oats. While participating in this scene there is a re-awakening of past lives, memories from a way of life that has all but vanished. . . . Cooking lessons with Maw Lang. The art of making a good cornpone. Use a small cast-iron skillet. Add shortening. Pour in batter. Cook on low heat until done. Turn over and brown on the other side. . . . The old variety of ribbon cane. The shorter sections at the bottom are the sweetest. Cane was fertilized with cotton-seed meal. Cut-back cane was used three years.

An October 5, 2005, sketch and journal entry about a fall crop of beans. I am always intrigued by the age-old methods my friends use in cultivating their crops. These methods have been used for generations.

under the pines. Children saved any spare change they might have earned working or won in competitions for fairway rides, toys, or candy.

Many stories I hear at the fair concern activities old-timers would enjoy during the fall months, and many of those tales involved hunting. Autumn's cooler temperatures bring migrating birds, waterfowl, and wildlife into the region's piney woods and swamps. Many farm families turned to market hunting as another way to earn a few extra dollars. Trading and selling meat and pelts helped families afford to stock up on winter supplies they couldn't make or grow. Many old-timers talked about capturing and driving wild turkeys and cattle toward the nearby Covington and Mandeville rail lines that headed northward, or south to New Orleans. At this time of year, wild duck, turkey, deer, and squirrel were plentiful, and they provided the main source of meat for farm families and were welcome additions to the table. Throughout the winter, farmers often disk plowed and baited their fields with cover crops to create a food plot that would attract turkeys and deer.

BRAISED WILD DUCK WITH GLAZED SWEET POTATOES

Makes 4 servings. Courtesy Dr. Jamar Melton.

———◇———

There are obviously numerous varieties of wild ducks, ranging in size from larger mallards to smaller teal. Wild duck has a flavor closer to beef than chicken, and tends to be lean and can easily dry out, so it is best to slow cook the meat with liquid. I like cooking the birds using juice from fresh Louisiana satsumas, a type of orange similar to a sweet tangerine. Many of the older cooks

I visit say you should begin by soaking ducks in cold water with two teaspoons each of salt and baking soda for about thirty minutes before preparing to cook.

- 2 mallards or 4 teal ducks
- 6 tablespoons salt, plus additional for seasoning
- 2 teaspoons baking soda
- 2 quarts water
- 2 teaspoons ground black pepper
- 2 tablespoons butter
- 4 small sweet potatoes, peeled and quartered
- 3 small apples, quartered and seeded, divided
- 3 small yellow onions, peeled and quartered, divided
- Peels from 3 satsumas
- 2 cups chicken broth
- 1 cup fresh satsuma or orange juice
- ½ cup Steen's cane syrup
- ¼ cup apple cider vinegar

1. Make sure all the feathers are removed from the ducks. In a glass bowl, combine salt and baking soda and add water to make a brine. Soak the ducks in the brine for 30 minutes. Remove from liquid and pat dry. Generously rub skins with salt and pepper.

2. Preheat oven to 275°F. In a large Dutch oven, heat butter over medium-high heat and brown the ducks on all sides. Remove ducks and add sweet potatoes and half of the apples and onions. Sauté over medium heat until lightly browned and covered with drippings.

3. Stuff the birds with remaining pieces of apples, onion, and satsuma peels. Place ducks back in the bottom of the pan, breast sides up, on top of sautéed apples, onions, and sweet potatoes.

4. Mix broth, satsuma juice, syrup, and vinegar and pour over the birds in the pan. Cover and bake for 2½ hours. Serve duck and potatoes warm with pan juices.

Final cash crops are harvested at fair time, and fall food crops produce up until the first frost, which might come as late as November. Many farmers used to plant a third crop of field peas in the fall, and they made sure to get their greens and onions in the ground once cooler temperatures set in. Even when the weather turned colder, fresh produce such as pumpkins, winter squash, and early greens were still staples. There were also crab apples, pears, and muscadine grapes to pick, eat, and can. Pecans, hickory, and ground nuts such as peanuts added texture and flavor to standards like pies, cakes, and candies.

1. Preheat oven to 325°F. Generously grease and flour a 10-inch tube pan. In a mixing bowl, combine flour, pecans, baking powder, and salt.

2. In a separate bowl, use an electric mixer on medium speed to cream together sugar and butter for 7 minutes. Use a low mixer speed to beat in eggs one at a time. Add vanilla.

3. With mixer still on low speed, add half the flour mixture. Mix in the buttermilk. Add remaining flour mixture and mix well. Spoon batter evenly in the prepared pan and bake until a knife comes out clean, about 1 hour and 15 minutes.

4. Remove from oven and cool 5 minutes. Flip pan over and place warm cake on a platter. Top with praline glaze.

PRALINE-GLAZED PECAN POUND CAKE

Makes 10 to 12 servings.

———◇———

This is one of those recipes that I searched for, for quite some time. A fair volunteer baked a batch of delicious pound cake similar to this recipe and passed it out one morning before the fair opened. This version created by Christina Melton is as close as it gets.

3 cups all-purpose flour

1 cup chopped pecans

1 teaspoon baking powder

½ teaspoon salt

3 cups sugar

1 pound (4 sticks) butter, softened

6 large eggs

2 teaspoons vanilla extract

1 cup buttermilk

Praline Glaze (recipe follows)

PRALINE GLAZE

Makes enough to cover one 10-inch cake.

———◇———

½ cup light brown sugar

¼ cup heavy cream

¼ teaspoon salt

2 tablespoons butter

¼ cup finely chopped, toasted pecans

1 teaspoon vanilla extract

1. Over medium heat combine brown sugar, cream, and salt. Bring to a boil and reduce heat.

2. Stir constantly until mixture thickens, about 5 minutes. Remove from heat and stir in butter until it is melted. Stir in pecans and vanilla. Let cool slightly to thicken. Spoon glaze over warm cake.

Peanuts and other ground nuts are a southern staple that came to America with enslaved Africans. They are boiled, roasted, parched, ground, stewed, or crushed for their oil. Many of my old friends recalled bringing peanuts in their lunch pails or carrying them in their pockets for snacks. To this day, a local favorite is "parched pinders," or roasted peanuts. I agree with the old-timers I talk with that taking fresh, unshelled peanuts and slowly roasting them in an iron skillet in an oven, preferably wood fired, gives them a rich flavor you can't buy. Boiling is another popular way to prepare peanuts. Just boil them in their shells for a little over three hours with a little salt and they are done. One of my favorite peanut varieties is the Pearl River County Peanut, from Pearl River County in Mississippi, which I got from Mr. Seldon Lang. He told me that this particular variety descended from an African American family from southern Mississippi that passed it on since the days of slavery. It's a light-pink skinned nut with a really sweet flavor. The hulls are kind of hard to crack, but the flavor is wonderful.

ROSE VISE'S NEVER-FAIL PEANUT BRITTLE

Makes about 1½ pounds candy.

———◇———

This homemade candy is one of the best uses for fresh peanuts. And it's as simple as it gets.

1½ cups white sugar

½ cup water

½ cup Steen's Cane Syrup or light corn syrup

2 cups shelled raw peanuts

½ teaspoon salt, plus extra for sprinkling

1½ teaspoons butter

½ teaspoon vanilla extract

1 teaspoon baking soda

1. Grease a cookie sheet or line it with foil. In a large cast-iron skillet put in sugar, water, and syrup. Bring to a boil over high heat. Stir occasionally with a wooden spoon until the mixture measures 300°F on a candy thermometer, usually between 10 and 12 minutes. At this point the mixture should "spin thread," meaning it has thickened to the point when drips from the spoon look like strands of thread.

2. Add peanuts and salt. Turn down to medium heat and stir constantly until mixture turns golden brown and peanuts have a "parched" look and smell.

3. Remove from heat and stir in butter and vanilla. Add baking soda and stir briskly until mixture foams up.

4. Pour onto prepared cookie sheet and quickly spread out with a large spoon. Sprinkle a little salt on top. Set to cool and break into pieces.

Occasionally I hear old-timers talk about the types of nuts they ate, such as goober peas and chufa nuts, or "chufies," which made healthy and tasty snacks. Goober peas are a type of ground nut, similar to peanuts, once grown in this region. Many people mistakenly associate goobers as being another name for peanuts, but they are not the same. Homer Graves told me that, when he was a boy in the 1930s, a local man named Lucius Crane grew a few rows of goobers. "They wuz roasted the same way you would peanuts," he remembered. "You could parch 'em in an ole arn skillet on the woodburning cook stove too and they had a good nutty taste to them." Goobers, or *Voandzeia subterranean*, were brought to America by enslaved Africans, and they were commonly grown before the Civil War. They were often ground into flour, pressed for oil, and used in the same way peanuts were used in later years. Chufa nuts are similar. Also originating in Africa, they are essentially a weed with a tuberous, almond-flavored ground nut that is high in protein. Both goobers and chufies were also used to feed hogs and attract turkeys. Both varieties are extremely rare and are not grown widely anymore.

———◇———

Stories that connect the people I meet with seeds that I collect are of particular interest to me, and nowhere is this idea better illustrated than at the fair. At the Mile Branch Settlement, for instance, you can sit in a rocking chair on the porch of the Pigott Cabin, where the family lived until around 1923, before sweeping changes overtook American agriculture. In terms of logs and construction, it is a physical piece of history. We have photographs of members of the Pigott family dating back to the 1830s. But we also have living artifacts that date back to that period, the Pigott field pea and Pigott red-cob corn. These seeds were actually grown by that family going back to the 1830s. It's just like the DNA of the family members themselves coming all the way down to the present, through the seeds they grew and that sustained them. We often like to say, "you are what you eat." In this case, we can actually consume these seeds as artifacts from an earlier time because they were cultivated by our ancestors. It is fascinating to consider that you can pass down family heirlooms and stories and study a community's history written down in a local library, but rarely can we say an actual living part of that history becomes a part of us and that we can play a role in passing that on. To me, as a farmer and a seed saver, that is something to claim and preserve.

Members of an older generation are good sources for seeds in any community, but so are local nurseries and

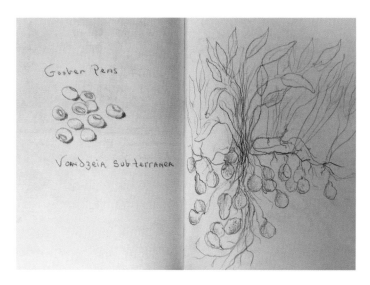

Sketch of goober peas.

farm-supply stores. One of my favorite places to visit in Washington Parish at fair time is the local Circle-T Feed and Seed Store on Main Street in Franklinton. I feel like a kid in a candy store when I am there. I have picked up a number of old varieties from the Circle-T that I grow up at Blackberry Farm. I also found a number of growers, like the late Mr. Red Fussell, who ran the local co-op, and Mrs. Margie Jenkins, born in 1921, who still operates a nursery in Amite, Louisiana, and is well-known for her rare, native azalea breeds. These local sources often have insight into the unique varieties people in the community are growing, along with a valuable level of expertise about how to cultivate local plants.

While I love hearing and recording stories and collecting seeds from people at the Mile Branch Settlement, my absolute favorite part of the fair has always been the agriculture exposition, where local growers showcase their best produce and unique, heirloom varieties. Judges looking for the most perfect examples of fruits and vegetables award cash prizes and ribbons for the

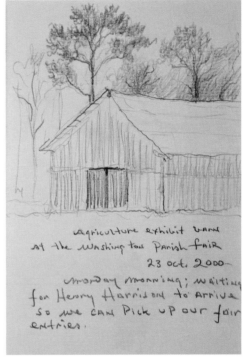

agriculture exhibit barn
at the washington Parish fair
23 oct. 2000
monday morning; waiting
for Henry Harrison to arrive
so we can pick up our fair
entries.

best specimens. For years, I would help bring entries to the contest from the farms of my older friends. I brought local varieties of beans, peas, peppers, onions, squash, and pumpkins. The best part of the growing season in this part of the country is that you can grow almost anything at a given time and season, so there is always a veritable cornucopia of produce that gets entered in the competition.

I always get excited when I see the entries in the fair's agriculture exhibits.

A 2000 sketch of the agricultural exposition building at the Washington Parish fairgrounds..

BAKED CUSHAW

Makes 10 servings. Courtesy of Mrs. Cornelia Weldon.

———◇———

One of the most intriguing vegetables I discovered at the Washington Parish Fair Agriculture Exposition is an old-fashioned crook-neck squash, locally known as a cushaw. Now cushaws have an illustrious history, having come to the American South from southern Mexico and nourishing the Native Americans who originally grew them here. This squash can grow to the astonishing weight of twenty pounds. It's a green-and-white mottled, striped crook-neck squash with a sweet, fragrant flesh. It repels pests, is easy to grow, and is good cooked in a myriad of dishes. Many people call it a cushaw pumpkin and, in at least one Creole cookbook from the turn of the last century, a pumpkin pie recipe called for tender cushaws by name. I have found cushaw seeds available through heirloom-seed companies. This recipe comes from the mother of Sarah Weldon Hackenberg, who took many of the photos for this book. It is delicious. Of course, if you can't find a cushaw, a sweet pie pumpkin will do.

1 medium cushaw

2 cups sugar

¼ pound (1 stick) butter, melted

2 large eggs

2 tablespoons all-purpose flour

1 teaspoon vanilla extract

½ teaspoon baking powder

1. Peel your cushaw. Cut it into quarters and scoop out the seeds. Cut the flesh into pieces and boil until tender.

2. Preheat oven to 350°F. Mash cooked cushaw and mix with remaining ingredients. Place in a 3-quart shallow baking dish and bake one hour. Serve warm.

The fair's agriculture exhibition also marks one of my greatest friendships in Washington Parish, with Henry Harrison, the longtime parish extension agent. Over the years, Henry has been a wealth of information. During fair time, he oversees the fair's renowned agriculture competition, and it is always thrilling to see what people enter and what wins. Henry was interested in agriculture from a young age and made it his career with the state's agricultural extension agency through LSU and Southern University. His life's work in supporting area farmers and educating the next generation is testament to the strength of Louisiana's 4-H program and, he believes, to the men and women who mentored him along the way. Growing up on his aunt and uncle's strawberry-and-produce farm in the Independence community in Tangipahoa Parish in the 1950s, Henry was surrounded by strong farming roots that bound the community together, including his childhood sweetheart and future wife's family, who have owned a local dairy farm in the community for more than a hundred years. Henry recalled how the schools he attended always had gardens tended by students who were encouraged to show off their produce at area farm competitions and fairs. One of his most influential mentors was a man named R. J. Courtney, who worked with small farmers through Louisiana's Community Rural Development (CRD) program for decades. With this strong foundation, Henry Harrison has continued to work with growers across the state of Louisiana, especially working with young, small farmers in Washington Parish.

Most recently Henry has focused on trying to bring younger farmers into production in the parish. "American agricultural production is a generational thing. The average age for the American farmer is over sixty-five years old," he told me on my last visit. "We need to find ways to encourage younger producers to get into

My friend Henry Harrison (*left*) works with farmers in the parish to support their businesses. He is pictured here with Seth Descant (*right*), a blueberry grower and farm coop operator in Washington Parish.

Sketches and journal entries from October 31, 2001, recording my departure from Washington Parish.

farming. When we talk about competing in this world and not being dependent, but being self-sufficient, agriculture and the ability to grow food is going to be the only thing that keeps us afloat in this country."

Like many of us, Henry believes the key to making family farming economically viable is to make it easier for small farmers to get their produce to market. One of the most promising developments is a program from the US Department of Agriculture that serves as a certification vehicle to verify that an independent grower uses "Good Agricultural Practices," or "GAP" for short. This type of certification allows larger supermarket chains, independent grocers, and restaurants to buy from local growers without the red tape. It is a voluntary program that offers training and documentation verifying safe and healthy growing practices, which provides confidence to these larger buyers.

Since World War II, farming has evolved toward large-scale agribusiness. Crops have been bred and genetically engineered to produce higher yields, to withstand higher doses of pesticides, and to have tougher skins and hulls that can be mechanically harvested, shipped longer distances to supermarkets, and have longer shelf lives. These practices resulted in increased output and less expensive produce, but they have also had the unfortunate result of breeding flavor and nutrients out of our food. High-volume mega-farms also squeezed smaller, experienced, independent farmers out of the market.

Just as a reduction in biodiversity of plant varieties is a threat to our food sources, so is the centralization of farming itself into the hands of a concentrated number of mass producers. Imagine what could happen if there was a major breakdown anywhere in the supply chain. Fortunately, local farms that are able to provide local residents with fresh, delicious, healthy food are experiencing a renaissance. Even the most secluded, rural areas now have thriving farmers' markets, farm stands, farm co-ops, or have developed what are known as "CSAs," or Community Supported Agriculture, which work like a subscription service for local farm produce. In many cases, chefs from area restaurants partner with growers to seek out and grow the flavorful varieties they want. I once had a famous visiting chef at Blackberry Farm tell me that, when he first started out, he was trained to use a lot of spices in his cooking to mask the flavors that were missing from the main ingredients. Thank goodness that has changed.

As valuable as commercial farmers are, smaller farmers and gardeners are the key to maintaining a diverse system of seed supplies and growers. However, many of these smaller farmers and gardeners also belong to an older generation, and they are dying off before passing on their seeds and expertise. If these seeds and traditions are going to survive, it's not going to be the US Department of Agriculture or the big seed companies that save them; it's going to be each one of you that takes this on. I don't care if it's just one tomato variety, or one variety of corn, or peanut, or pea. You may be the only person that's taking care of that particular variety and the cultural legacy that surrounds it, and that alone can save it.

I really believe the growing revival in farming and farm-to-table eating we are experiencing is fueled by people's desire to know what they are eating and where their food comes from, and to reclaim the flavor of some of these older, heritage varieties. Not only are they tasty and pretty, but they're higher in vitamins and nutrients.

This renewed interest in the older varieties has also sparked interest in some of these age-old farming techniques that we call "organic," but that evolved as a matter of necessity. Many of these techniques involve rotation of cops to maintain soil health and reduce the use of fertilizer. These practices rely on biodiversity and interplanting flowers, herbs, and cover crops to create an environment that improves pollination rates, reduces crop-damaging pests, and decreases the need for pesticides. For example, certain plants attract bees and beneficial parasitic wasps that pollinate crops and kill cabbage worms, grasshoppers, tomato hornworms, two-spotted stinkbugs, and Mexican bean beetles. We know that many of these older varieties have adapted to local conditions where commercially available seeds have not. It is this type of adaptation and accumulated agricultural knowledge that's going to help protect us against future threats to our food supply, if only we are wise enough to save the legacy we've been given.

As fair week comes to a close, before I pack and head home, I always feel a tinge of anxiety and a desire to make the most of my remaining moments in Washington Parish. Each year, I know there is a chance that these cherished visits with my good friends might be my last chance to talk with them. While I make new friends every time I attend, and I record more stories and information, I find fewer and fewer familiar faces. I still make lists to remind myself about questions to ask, and about the seeds to gather, share, and plant, but there is never enough time.

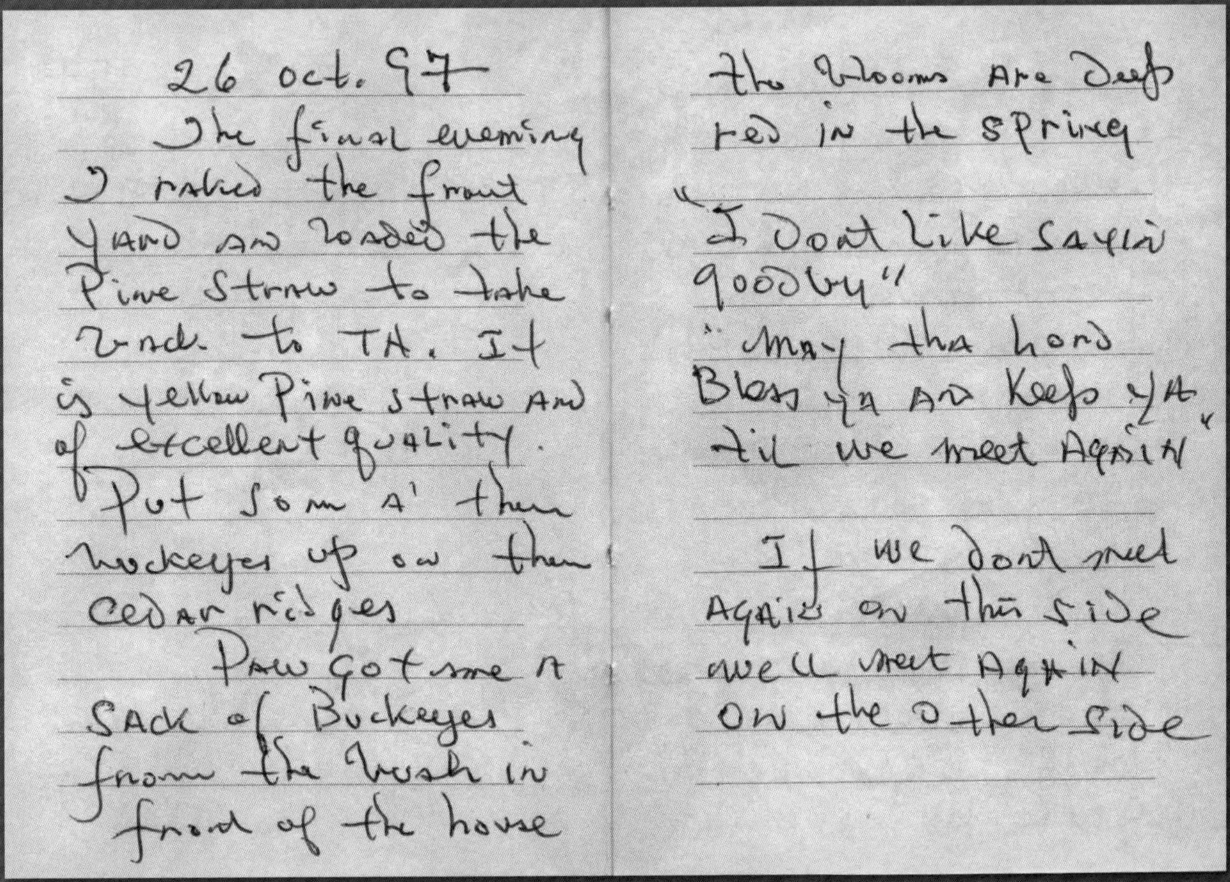

26 oct. 97

The final evening I raked the front yard an loaded the Pine Straw to take back to TN. It is yellow Pine straw and of excellent quality. Put som a' them buckeyes up on them Cedar ridges

Paw got some a sack of Buckeyes from the bush in frond of the house

tho Blooms Are Deep red in the spring

"I Dont Like sayin goodby"

"May tha Lord Bless ya An Keep ya til we meet Again"

If we dont meet Again on this side we'll meet Again on the other side

A journal entry from October 26, 1997, detailing my last evening in Washington Parish before the trek back to Tennessee.

STARTING AGAIN

On a recent trip to Louisiana, my friends got together to recognize my years of service volunteering at the Mile Branch Settlement of the Washington Parish Fair. It was a truly gratifying experience. My friend Gus Magee even made a special rocking chair for me with my name carved into it. It will serve as a daily reminder of my treasured relationships when I am sitting at home in Tennessee.

As the year passes and another one begins, I find myself reflecting on the passage of another series of seasons, yet again. As I think ahead to my future trips to Washington Parish, I know that what keeps me coming back is a special feeling of home with people that I love, people who share my passion for farming and traditional ways of growing things. The people, the land, the seeds, and the stories meld together to make something that is irresistible. With each new visit, I will continue recording stories as long as I am "on that green side of the turf," as Mr. Homer would say. As much as life and family in Tennessee have shaped the person I am, my life would be incomplete without my visits to Louisiana. They are as integral to the person I am as eating and breathing. I can't say that I've made much of a contribution to life in Washington Parish, but life here sure has contributed to me, and I am eternally grateful for that.

Every year, I reflect on my final moments in the par-ish before I return home. As I wrote in a journal from 2006, these moments are often filled with a sense of divine creative inspiration: "Never before have I been so thoroughly driven as if by outside forces to complete my work here. I have worked with pen writing and with pencil drawing, almost with a frantic pitch, attempting to capture and preserve for all time each relevant scene, story, and piece of history before they fade and then forever vanish from our lives. The smell of strong black coffee, bacon frying, and biscuits baking in the oven fire in the kitchen. Mr. Seldon and I are seated at our usual place at the kitchen table. As always, my notebook is open and pen poised, waiting for the instant when a story surfaces from his seemingly endless collection of life's experiences."

Like fading recollections and images, I see the lives of my old friends passing on before me. There is a sense of urgency to capture their history, stories, knowledge,

I am th same today, yesterday, tomorrow and forever. "Be rooted and grounded"

Final words of wisdom recorded from my friends,
as I depart for home.

and wisdom before it is forever gone, no more to be
known or available. With their passing, this place and
our lives will be much poorer. As in the past, I will always
be hanging on to the last possible minute, waiting for
one last priceless gem of a tale to come to the surface
before it is time to depart.

RECIPE INDEX

Baked Cushaw, 183

Beans and Peas

    Beulah Mae Lang's Butter Beans, 102

    English Peas and New Potatoes, 96

    Field Peas or Lady Creamer Peas, 67

    Hoppin' John, 103

    Maw Lang's English Pea and Tomato
      Salad, 130

    Summer Succotash, 130–31

Beulah Mae Lang's Biscuits, 103

Beulah Mae Lang's Butter Beans, 102

Beulah Mae Lang's Tea Cakes, 111

Beulah Mae Lang's Tomato Gravy, 75

Biscuits, Beulah Mae Lang's, 103

Blueberry Peach Cobbler, Rose Vise's,
    146–47

Braised Wild Duck with Glazed Sweet
    Potatoes, 175

Bread and Biscuits. *See also* Cornbread
    and Cornpone

    Beulah Mae Lang's Biscuits, 103

    Dumplings (flour dough), 142

Butter Beans, Beulah Mae Lang's, 102

Butter, Mason Jar, 63

Cabbage, Smothered, 168

Cake Glaze, 144

Cakes

    Beulah Mae Lang's Tea Cakes, 111

    Coconut Pound Cake, 144

    Praline-Glazed Pound Cake, 176

    Syrup Cake, 76

Candy

    Rose Vise's Never-Fail Peanut
      Brittle, 178

Chicken

    Chicken and Okra Stew, 168

    Chicken Pie, 143

    Lang Family Chicken and
      Dumplings, 142

Cobbler, Rose Vise's Blueberry Peach, 146–47

Coconut Pound Cake, 144

Collard Greens, 56

Corn

    Skillet Corn, 133

    Summer Succotash, 130

Cornbread and Cornpone

    Cornbread Dumplings, 58

    Fried Cornmeal Dumplings, 110

    Lighter Cornpone, 61

    Ruthie Mae Graves's Cornbread, 63

    Ruthie Mae Graves's Turnips and Greens
      with Cornbread Dumplings, 58

    Traditional Simple Cornpone, 61

Creamer Peas, Lady, or Field Peas, 67

Cucumbers

    Refrigerator Dill Pickles, 134

Cushaw, Baked, 183

Desserts

    Beulah Mae Lang's Tea Cakes, 111

    Coconut Pound Cake, 144

    Praline-Glazed Pound Cake, 176

    Rose Vise's Blueberry Peach Cobbler,
      146–47

    Rose Vise's Never-Fail Peanut
      Brittle, 178

    Rose Vise's Sweet Potato Pie, 73

    Syrup Cake, 76

Dill Pickles, Refrigerator, 134

Duck, Braised Wild, with Glazed Sweet
    Potatoes, 175

Dumplings

    Cornbread Dumplings, 58

    Dumplings (flour), 142

    Fried Cornmeal Dumplings, 110

    Lang Family Chicken and
      Dumplings, 142

Eggs, Stuffed, Rose Vise's, 105

English Peas and New Potatoes, 96

English Pea and Tomato Salad, Maw
   Lang's, 130

Field Peas or Lady Creamer Peas, 67
Fig Preserves, 124
Fish, Pan-Fried, 108
Fried Cornmeal Dumplings, 110
Fried Okra, 131
Fried Potatoes, Skillet, 169

Glazed Sweet Potatoes, Braised Wild
   Duck with, 175
Glazes for Cakes
   Cake Glaze, 144
   Praline Glaze, 176
Gravy, Beulah Mae Lang's Tomato, 75
Greens
   Collard Greens, 56
   Ruthie Mae Graves's Turnips and Greens
      with Cornbread Dumplings, 58

Homemade Mayonnaise, 130
Hoppin' John, 103

Kilt Salad, 94

Lady Creamer Peas, or Field Peas, 67
Lang Family Chicken and Dumplings, 142
Lettuce
   Kilt Salad, 94
Lighter Cornpone, 61

Mason Jar Butter, 63
Maw Lang's English Pea and Tomato
   Salad, 130
Mayonnaise, Homemade, 130

Never-Fail Peanut Brittle, Rose Vise's, 178
New Potatoes, English Peas and, 96

Okra
   Chicken and Okra Stew, 168
   Fried Okra, 131
   Summer Succotash, 130–31

Pan-Fried Fish, 108
Peach Cobbler, Blueberry, Rose Vise's,
   146–47
Peanut Brittle, Rose Vise's Never-Fail, 178
Peas. See Beans and Peas
Pecans
   Praline-Glazed Pound Cake, 176
   Praline Glaze, 176
Pickles, Refrigerator Dill, 134
Pies
   Chicken Pie, 143
   Rose Vise's Sweet Potato Pie, 73
Potatoes
   English Peas and New Potatoes, 96
   Skillet Fried Potatoes, 169
Pound Cakes
   Coconut Pound Cake, 144
   Praline-Glazed Pound Cake, 176
Praline Glaze, 176
Preserves
   Fig Preserves, 124
   Strawberry Preserves, 100

Refrigerator Dill Pickles, 134
Rice
   Hoppin' John, 103
Rose Vise's Blueberry Peach Cobbler,
   146–47
Rose Vise's Never-Fail Peanut Brittle, 178
Rose Vise's Stuffed Eggs, 105
Rose Vise's Sweet Potato Pie, 73
Ruthie Mae Graves's Cornbread, 63
Ruthie Mae Graves's Turnips and Greens
   with Cornbread Dumplings, 58

Salads
   Kilt Salad, 94
   Maw Lang's English Pea and Tomato
      Salad, 130
Simple Cornpone, Traditional, 61
Skillet Corn, 133
Skillet Fried Potatoes, 169
Smothered Cabbage, 168
Squash
   Baked Cushaw, 183
Stew, Chicken and Okra, 168
Strawberry Preserves, 100
Stuffed Eggs, Rose Vise's, 105
Summer Succotash, 130–31
Sweet Potatoes
   Braised Wild Duck with Glazed Sweet
      Potatoes, 175
Rose Vise's Sweet Potato Pie, 73
Syrup Cake, 76

Tea Cakes, Beulah Mae Lang's, 111
Tomatoes
   Beulah Mae Lang's Tomato Gravy, 75
   Chicken and Okra Stew, 168
   Maw Lang's English Pea and Tomato
      Salad, 130
   Summer Succotash, 130–31
Traditional Simple Cornpone, 61
Turnips and Greens with Cornbread
   Dumplings, Ruthie Mae Graves's, 58

Wild Duck, Braised, with Glazed Sweet
   Potatoes, 175

Table of Contents

Table of Contents   1

Concerning Thanks …   2

Concerning Beginning…   5

Concerning Eden…   14

Concerning Gardens …   27

Concerning Gardening …   46

Concerning Garden Keeper …   78

Concerning Soil…   102

Concerning Seeds …   126

Concerning Weeds …   148

Concerning Compost …   168

Concerning Stones …   192

Concerning The End…Before We Begin.   210

Concerning Thanks . . .

For Jeff: my husband, my sanctuary, my fountain
Songs of Songs 4:12

For my parents: who planted, weeded and guarded me.
Thank you for the shovel that can't be broken.

For my sisters: who taught me to master chaos and beauty.
For my brothers: who taught me to love playing in the dirt and made me brave.

For Susan: who taught me to garden
For Stacy: who watered me
For Lorry, Cindy, Amanda, Pastor Keith, Pastor Mike and the garden fairies of St. Paul's Garden Circle:
Whose lives taught me to love the Gardens of the Lord.
For the Johnstown Writers Guild and Betty Rosian

"...like the brightness after the rain that brings the grass from the earth." 2 Samuel 23:4b

For little Bug and my bean: who taught me more than I ever taught them and made my life full.
Please remember this much about God's plan for you.

Thank you for the wings.

"*Forever is composed of nows.*"
– Emily Dickinson

Concerning Beginning. . .

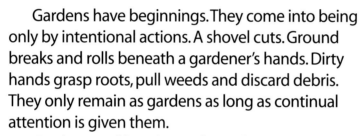

Gardens have beginnings. They come into being only by intentional actions. A shovel cuts. Ground breaks and rolls beneath a gardener's hands. Dirty hands grasp roots, pull weeds and discard debris. They only remain as gardens as long as continual attention is given them.

Gardens are like relationships. They begin with proximity. A single seed, like a single word, plants something beautiful or the seedling of something that may disrupt beauty. Intentional kind attention cultivates fruit, flower, or a labyrinth of wonder for the soul. Neglect, harsh actions or apathy far more easily destroy them.

Relationships are like gardens. Neglected landscapes of humanity map the globe. Yet human lives always cultivate something. Proximity rolls the soil. Attention plants seeds. Neglect harbors roots that silently destroy. Contact increases bonds or harbors a maze anchored in chaos.

This journal is an account of how the Lord pursued me (one He loved) through something He created. It is a collection of stories and Biblical parallels recorded to share how He has drawn me to Himself.

At a young age, a love for the garden and gardening took root within me. My dad built my sisters and me a table with lights in our basement. We used it to plant seeds early in the spring. Mom bought seed packets of her favorite flowers. We planted pansies, hollyhocks and columbine. She told stories about the people each of those flowers brought to mind. My own memories added to that collection. Portulacas remind me of grandma. We planted them, and their sturdy beauty greeted me every time I visited. I remember my brother buying a Lupine for mom. That flower ties that gift to itself every time I see a Lupine. Hydrangeas fill my mind with vibrant stories of an old farmhouse. I love those memories. I loved those seeds. Treasure took root and has been growing since.

Cut lilacs always filled vases and peonies dropping petals on the counters triggered my constant attention. I remember my sisters and I cutting crabapple branches covered in blossoms. We formed them into wreathed centerpieces on our dining room table. I wanted dad to delight in them. One Easter a green inch worm stood on its hind legs to reach up for a guest's plate, and when I remember him I still hear the laughter. Brave soul crawled out of his water bowl and inched across the table amongst warm bowls sliding among six hungry children. With so many diversions I never conceived that so much more than wreathed flowers was forming in my heart during these growing years.

Adversity's blessings are often hard to see. Gardens come with weeds, bugs, and thorns. Turning the soil and ripping away weeds a sweaty mess. Little agitations shift, shake and rouse emotion. Gifts of all shapes and sizes grown only from the distance time provides. Sun-brewed tea dancing over ice and the sweet garden strawberries over melting ice cream waiting for my brothers and me when we came in reinforced what the work could produce. It was unveiled in tired laughter between yummy bites. I used to doubt that mom knew how special those gentle kindnesses became in each heart they touched. But now that I am a mom, I know she knew. Our family fought with each other, but made time to gather apples. Bonfires were built with broken branches. We warmed ourselves side by side safe in the refuge. We shoveled though, weeded out and planted the gardens within each other. Each memory planted seeds, cultivated the soil of my soul and sent roots into the deep dark of what made me a gardener.

Soon after I married, my parents sold the farm we shared as a family. Dad gave me an old shovel. It was part of well-used, yet no longer needed, tools from my childhood farm. It is even more battered since it has been in my hands, but it was worn from decades of use long before it was mine. It is stainless steel from tip to top. It is one of my favorite things, even though he has given me other more valuable things. I am sure the shovel was not intended as a gift. But I think, perhaps I became a gardener because dad gave me the tool to be one. The excellence of some gifts is hidden in the inheritance we cultivate from them not the value we identify in them. Without bending or breaking, it dug the soil of every garden I have started and a hole for every plant I have transplanted since.

I love being a gardener. In a world of physical and metaphorical ruins, gardens bud with hope. There will always be desert wastelands where fruit is devoured as branches are striped and roots are clawed. As the seasons of life change, thorns and drought will be as common as grief and loneliness. It is good to know that evergreen forests stand despite frigid temperatures, but it's just as important to comprehend survival in such places. I understand that jungles define chaos and that we all feel lost until we put down roots. Despite floods that drown, the paths of tornadoes, and winter's icy blanket, gardening provides the tools needed to carry on despite loss.

I am from the Laurel Highlands of Pennsylvania. The green woodlands have an ancient beauty that giggles as streams roll through the hills. The fall is beautiful, but winter comes every year. This climate trains a person to be ready for constant change and unpredictable conditions. Gardening through the changes promises new hopes every morning.

"The Lord's loving kindnesses indeed never cease,
For His compassions never fail. They are new every morning;
Great is Your faithfulness." Lamentations 3:22-23

My observations of these truths help me to see my life and the lives of people around me very differently. Ugly things can be pulled away with our hands and beautiful things can be gathered in our arms. Believing in the Creator's hand in all of it helps me survive the chaos. Glimpses of His handiwork whisper His name, I look for them. He is in the song of peepers at night and bird's song in the morning, I listen.

Seeing myself as a gardener has helped me understand my role in a struggling world. Answers to nameless longings and questions I didn't know how to ask I found in the simple truths of a garden. Weeds grow in the same places as fruit and blooms. When something or someone is wilting, water and attention aid in its survival. Not every plant or person flourishes in the full light of day. Like shade gardens, some of the most pleasant people and places are hidden away in sheltering havens. Growth is important. However, pruning or transplant is necessary when the stems of plants and hopes of the heart struggle to reach for light. If we want to eat fruits and vegetables, patience is vital. We have to wait for growth and allow the fruit time to ripen. The same patience is required with children and friendships. The rules that govern gardens are unambiguous. Despite all the ages of the earth, we have yet to understand all that the Lord teaches through them.

"This is what the Lord says; he who made the earth, the Lord who formed it and established it" The Lord is his name. Call to me and I will answer you and tell you great and unsearchable things you do not know." Jeremiah 33:2-3

Each part of the garden communicates diverse truths in different times of the year. In different stages of life, a person interprets each differently as well. A garden teaches in a unique way. Each part of the world reminds us to give credence to garden wisdom: observed, touched, smelled and tasted. The seeds, roots and flowers His wisdom formed remind us not to forget.

"By wisdom the Lord laid the earth's foundations, by understanding He set the heavens in place; by his knowledge the deeps were divided, and the clouds let drop the dew." Proverbs 3:19-20

This collection of observations would not be complete without a guide for interpretation. For that reason verses that have opened my eyes, fed my soul or dug up questions are included in this journal. Evidence of His love cascades from passages and promises in the Bible. Each of them applies to our lives in countless new ways. I studied as I read and ask the Lord to open my eyes to hidden truths.

"I applied my heart to what I observed and learned a lesson from what I saw." Proverbs 24:32

While choosing to share the insight I've gathered, I know it doesn't apply to every situation, as money can't buy wealth or a big heart love. But gardeners grow their own treasures. As Solomon hunted for hidden treasure in Proverbs 2:2-5, I searched each verse over and over, and believe every word. I begged for understanding. I sorted through the ones about gardening or about God's love and added constantly to this collection. They are scattered here like seeds to plant, encourage others, and use as a guide.

While, "turning my ears to wisdom, applying my heart to understanding and crying aloud for insight ..." God's goodness passes before us. In every sunrise, flower, or waterfall we get to see our value to him. He covers us with His hand and reveals his glory. Hidden treasure is buried in everyone we meet (shallow and deep). Complicated as people are, they are worth watering, cultivating, digging beneath their crusts, trusting seeds.

It is my prayer that the truths recorded here will be a drink for the wilting, warmth for those in the middle of winter and light for those living in darkness. May it be fertilizer for those in bud, in full bloom, or ready to fruit.

"The Lord will surely comfort (His people, me) and will look with compassion on all her ruins. He will make her deserts like Eden, her wastelands like the garden of the Lord. Joy and gladness will be found in her, thanksgiving and the sound of singing." Isaiah 51:3

GARDEN
Always

Concerning Eden . . .

"The Lord God said, 'It is not good for the man to be alone. I will make a helper suitable for him.'" Genesis 2:18

It wasn't wrong for Adam to need a relationship with someone other than God. A person can change a place. In the same way a place can change a person. Like waterfalls that surprise someone lost in a jungle or a sunset we meet at the crest of a mountain, there are souls we meet that can change everything. We are designed for relationships. A rare soul helps us see the stars through the darkness and then the aurora borealis dancing across the sky. All of us search for moments in such places, even if we only know they exist because of stories told of them.

"Deep calls too deep in the roar of your waterfalls; all your waves and breakers have swept over me." Psalm 42:7

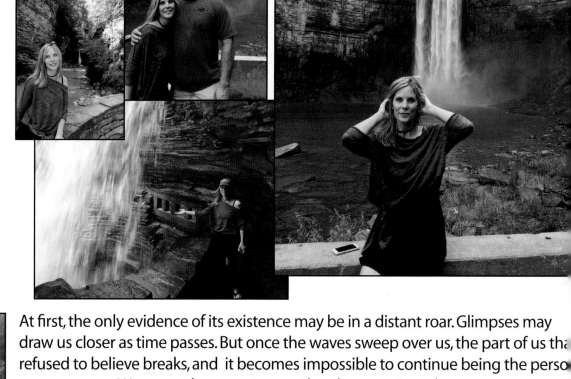

At first, the only evidence of its existence may be in a distant roar. Glimpses may draw us closer as time passes. But once the waves sweep over us, the part of us that refused to believe breaks, and it becomes impossible to continue being the perso we once were. We can no longer return to the place we started.

Even as a teenager I caught glimpses of Eden, though much of what constitute a garden was infuriating. For example, planting my first batch of seedlings was a disaster. I took trays of the little green starters outside and worked in the yard nearby. I thought, the seeds would soak up spring. Grand flowers bloomed and cascaded through my hopes as I prepared a place for them. At the end of the morning's work, I decimated all the seedlings. When I went to check, the sun baked them. I had yet to understand why or how my mother hardened them off, by gent introducing them to the harsh realities of the real world. There was no way to resto them. I lost everything for which I had waited. Mom and my sisters were mad. I was heartbroken. It was maddening to think of what was lost in one bad decision.

I loved being outside. Confined indoors? Not so much. I did not like canning tomato sauce, jellies, or pickles as a teen. There were too many bosses. All of my sisters, Grandma and Mom had too many opinions. That is remarkable, because now I look forward to making sauces and jams in late summer. I gather fruit and canning supplies. I put in old movies and watch them while trapped all day. But I would give anything to spend a day with my family canning again. Of course, I would have them cut the bad spots out of the tomatoes and strawberries. I would taste the blueberries and stir the warm pots on the stove. I would breathe in the fragrances and soak up my family. I would quietly listen as they all shared stories or argued about differences and love every minute of it. Grandma died since and my sisters and mom have all moved. My five siblings and our families rarely all get together, even for holidays.

As restoring lost seedlings or gathering loved ones from the grave is outside our strength, so Eden is beyond our hope. Yet it surprises me that much of what infuriated me then, I miss so much now. I guess we all have moments and places we wish we could have back. I think our whole lives our hearts are awake, but we sleep.

"I slept but my heart was awake. Listen! My lover is knocking . . ."
Song of Songs 5:2a

Life gives us tastes of happiness that we do not always recognize as we breathe and swallow them. But years later, we savor any activity that reminds us of them. Life is a search for an endless love that has always been ours. Some things can never be found. "The eye never has enough of seeing, nor the ear of hearing." Ecclesiastes 1:8b. The heart never has enough of longing. We dream of a promised land, talk about heaven and remember Eden, because we all seek a place we believe exists, perhaps because we have seen glimpses of it in the sun breaking through the darkness.

"He is like the light of the morning at sunrise on a cloudless morning, like the brightness after the rain that brings the grass from the earth." 2 Samuel 23:4

We look for someone to love through all of its jungles, as the girl in

Song of Solomon 3:2. "I will search for the one my heart loves. So I looked for Him but did not find Him."

Yet all the while we are surrounded by His kindnesses and guarded by His love.

Throughout our lives we look for a place we have never been though it is familiar. We don't remember when the longing began. But the hope for it creeps into each child's dreams and does not diminish. It is a place filled with the simple joy we felt as we ran and jumped into the sky reflected in puddles as children. The memory of it fades as we grow, but the longing creeps up again and again in quiet moments. In every foreign destination we explore and each vacation we take, we hope to find it. When age has memorized every path we walk, it will be the place for which we still look around every corner. It is the place we are reminded of when our fingertips touch a moss covered rock. Even the sound of water and the smell of swee peas or jasmine along a garden wall draw up faint memories just o of our reach of a place where we belong and that owns us. The late evening sun reflecting on water's surface strengthens our hope. A place closer to us in moments of joy, it quickly fades to something hopeless again. Renewal and intimacy have to be its purpose. Is it a place we can never reach? Yet we recognize it. We can imagine it. Though its unimaginable beauty exists only in our hearts, it is yet a perfect place, not imaginary. It was a garden. It was first named and still called Eden. It is yet guarded by its unforgettable nature.

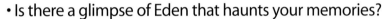

- Is there a glimpse of Eden that haunts your memories?
- Is there a soul who changed the world for you?
- Is there some trying part of your childhood you now recognize was important to enjoying life now?
- What longing keeps you awake while the world sleeps around you?
- Has God ever revealed Himself to be the answer through that longing?
- Are there places you go? Things you long to see, smell, touch that stir up hope within you?

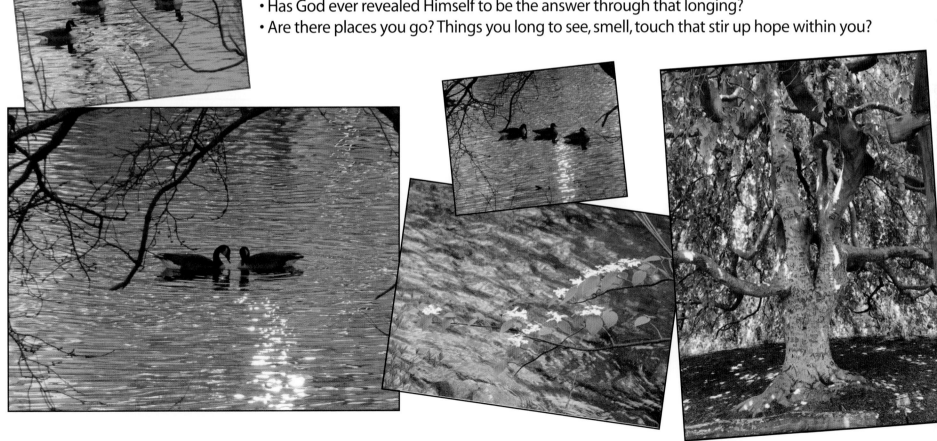

Longing . . .

In the expanse of space and time that filled Adam's heart to the time that now fills ours, a longing for our Creator has not faded. We stand at two ends of the earth. Yet from the Garden of Eden to the garden in my backyard, God's love has not dried up in any drought or been overrun by countless weeds.

"Those who live at the ends of the earth stand in awe of your wonders. From where the sun rises to where it sets, you inspire shouts of joy." Psalms 65:8

The longing for that meeting place with a god worth worshiping has not diminished or been forgotten in all the time and distance that separates Eden's first occupants and us. David recorded the same desire in Psalms 42.

"As the deer pants for streams of water, so my soul pants for you, O God. My soul thirsts for God, for the living God. When can I go and meet with God?" Psalms 42:1-2

The desire for a haven from the world, oasis in desert and hope for a promised land evokes all that Eden's name conjures. Those desires mark a map for our soul; a map with directions where we can meet Eden's creator and the answer.

"He has made everything beautiful in its time. He has also set eternity in the hearts of men; yet they cannot fathom what God has done from the beginning of time." Ecclesiastes 3:11

God is still speaking through the place He originally designed for and spent time with man. Designed countless years ago, gardens are special today and will be for unknown years to come because of the eternal answers they echo. The evidence of our Creator's infinite love has not been withheld or hidden. Maybe just overlooked and taken for granted.

Gardens can't spin like a compass. Neither does a garden purposely point north, south, east or west. What they offer provides a universal picture. Constant, gardens are never the same, always changing. The Lord uses them to herald this truth.

The garden's intricate treasures and bold focal points honor the Creator, His power and gentleness. They are growing, fruiting, blooming things that remind man of His infinite wonders. The colors, fragrances, and details are for enjoyment. Gardening then becomes a gift and maybe an act of worship. Individual acts enduring, because they scatter seeds. Their cotyledons pushed through the darkness and their roots send out runners that are still running.

• Have you seen God's love echoing through a garden?

• Where else in life do you suppose His love is overlooked and yet extravagantly poured out?

Adam and Eden...

"... the Lord God formed the man from the dust of the ground and breathed life into him, the breath of life, and the man became a living being. Now the Lord God had planted a garden in the east, in Eden; and there he put the man he had formed." Genesis 2:8

While I have heard the story of Adam and Eve countless times, I only recently observed the story of Adam and Eden as I began to journal. In Genesis 2:8, notice that God created a garden for Adam before He created Adam! He created a garden and then a gardener for it. He made us gardeners before sin entered. God's purpose for our lives has not changed.

"I know that everything God does will endure forever; nothing can be added to it and nothing taken from it. God does this so that men will revere Him." Ecclesiastes 3:14

We need the Lord no more, no less than Adam needed Him. But we need to work harder to open our hearts to Him and know His thoughts. This truth has not changed since the first bite of sin. Our lives are no longer connected to the soil as Adam's and countless generations of gardeners and farmers since have been. We go to the store or order food online. But we were created to care for things that flower, fruit and bear seeds. Life's underlying purpose has not changed.

All gardens are not made up of soil and plant life, but our lives always cultivate something. Every life sends out roots, plants seeds, and impacts life around it. Every day there are things and people that need water or that are drowning in too much. Sometimes there is not enough sunshine, and other times something or someone withers from too much exposure. Both figuratively and physically, water and light must be present. The Bible tells us in Amos 4, God has His reasons for struggle.

"I also withheld rain from you when the harvest was still three months away. I sent rain in one town, but withheld it from another. One field had rain; another had none and dried up. People staggered from town to town for water but did not get enough to drink, yet you have not returned to me." Amos 4:7-8

Our Creator makes Himself known in a gardener's arduous labor and quiet of a gardens enjoyment. And in both we find truth and beauty. He is the key to the map. Through Him, we come to recognize what is beneficial and what is destructive. Through Him, we appreciate more than seedtime and harvest.

Gardens have specific needs: water and lighting, soil pH, temperature zoning, and so much more. Understanding those needs takes study. After that, the battle has only just begun. Both good and bad plants spit seeds that germinate where they are not planted. Good root systems are cut through and overrun by the roots of iniquitous growth. Troublesome undergrowth sends out vines that strangle and slowly strip away light and moisture.

The simple truths of a garden apply to a person's relationships and life investments. If the needs of vegetation are complicated, the people we share gardens with are infinitely more complex and important to understand. Each individual has lighting and water requirements, pH tolerance, and temperature range. We are surrounded by things and people, who may be perennially a part of our lives, or just annuals, here for only one season. Enjoying them while they are here and letting them go at the right time a needed skill. Some lives will tower over us and others we must be careful not to smash under our feet. None should be taken for granted.

Whether our lives produce a garden, a jungle, or a desert; blessing or burden depends on if and how it is tended. A cultivated life can bloom with its own whispers of Eden, if time is taken to nurture it.

• Do you feel that God created a place for you before He created you? A purpose for your life that cannot be changed?

• Have you ever had the wisdom to care for a person scorched by exposure, drowning in trouble or withering in darkness?

• Has a short lived relationship (annual) ever impacted you as much as a perennial one?

• Are you more careful with relationships that tower over you or ones that can be trampled easily?

Fallen Nature...

I struggled with nature's ability to teach when I began journaling. The ways of nature can be very undependable. Forget the garden. Look at the ocean. It can sit calmly and reflect every star above it. It can magnify the sunrise and sunset in its mirrors. The rolling waves along the shore entrance men and women who live beside them every day and those who visit it once in a lifetime. Yet a little wind can sicken a boatman and strong winds over it can swallow a ship. Warm temperatures and wind both can decimate land miles from shore. Tornadoes, drought and wildfires come and go in unpredictable successions.

Fallen nature can obscure God's hand. C.S. Lewis said ... nature cannot be used as a teacher. It will always lead to dark gods or a great deal of nonsense. I agree with him.
I asked my pastor what he thought. He pointed out Roman's 1:19-20.

" Since what was made known about God is plain . . . because God has made it plain . . . For since the creation of the world God's invisible qualities – His eternal power and divine nature – have been clearly seen, being understood from what has been made, so that men are without excuse."

God knew chaos would come and that the world would need care and gardened. He knew man's basic need for food and deep-seated need for beauty and order. So the first thing He taught man was to take care for it in a place we would always need. He introduced man to a sanctuary that would always whisper of His power and, if cared for, could herald His glory.

"I have seen you in the sanctuary and beheld your power and glory." Psalms 63:2

Gardens have power to send roots deep in our souls. They have a way of becoming friends. Though silent, they never stop talking. They pull us toward gentleness and goodness and change us when we allow them.

C.S. Lewis said this about God's reason for becoming a man.

"God became man to turn creatures into sons; not simply to produce better men of the old kind but to produce a new kind of man. It is not like teaching a horse to jump better and better but like turning a horse into a winged creature."

He gave us gardens to remind us we are more than we appear. Gardening has the power to give us wings in this fallen world.

I imagine when God met Adam in Eden, He taught him what kept a garden healthy. He would have shown Adam how to work with the soil, plant each seed and tend each plant. I imagine they wondered together about this detail and that, and that God allowed Adam to figure some things out for himself. How we meet God has changed, and what constitutes a garden, at least one that functions as Eden, has changed greatly as well. But I have seen more than a few gardeners sprout wings under a garden's magic! As Fredrick Buchner said, "The place where God calls you is the place where your deep gladness and the world's deep hunger coincide."

God still teaches us all we are willing to learn through the garden. God does not tutor us to garden today as He did Adam, but He gives us endless reasons to collect and cultivate all that He has given. His kindnesses are evident if we look for them. John testifies to this in the last sentence in the book of John.

"This is the disciple who testifies to these things and who wrote them down. We know that his testimony is true. Jesus did many other things as well. If every one of them were written down, I suppose that even the whole world would not have room for the books that would be written."
John 21:25

Since discovering this truth about a garden, I look at my life as a garden. Whether I am at the edge of the sea, the foot of a mountain or have my face deep in a bouquet of lilacs, I search for the places in the world that speak to me about Him. I worship in those moments and meet Him in my prayers. I make decisions as though I am gardening and consider myself a gardener. I try to be intentional about everything I do. Mundane tasks make more sense when I see them through God's first purpose for any man's life. Remembering good things and focusing my efforts on what can be restored or made better makes every day purposeful. Independently, no undertaking may look like it is ambitious or important. But all of the little tasks put together have tremendous purpose. They cultivate a place worth celebrating with fruit worth sharing and flowers worth collecting. And so I garden always.

• Is there a sanctuary, other than your church, that you run to get your mind off the world and focus on God?

• Do you feel that any part of a your life reminds you that God wanted more for you than the burdens of a fallen world?

• If you could walk through Eden with the Lord and were not burdened by sin, what would you share with Him and desire God to reveal to you?

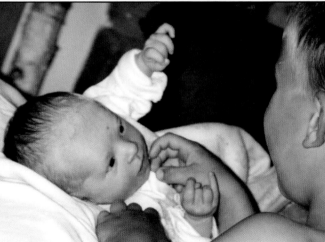

GARDEN
Always

Concerning Gardens. . .

"About Joseph he said: 'May the Lord bless his land with the precious dew from heaven above and with the deep waters that lie below; with the best the sun brings forth and the finest the moon can yield; with the choicest gifts of the ancient mountains and the fruitfulness of the everlasting hills; with the best gifts of the earth and its fullness and the favor of Him who dwelt in the burning bush . . ."
Deuteronomy 33:13-16 (Joseph's blessing)

What a blessing to wish over a child's life! The land blessed by one generation has tremendous power to bless the next. The gardens around our old farmhouse remind me of such a blessing. Before it was our home, a perfect stranger gathered dew and dug a well. He or she harnessed the best the sun could bring forth and the finest the moon could yield. A predecessor left favorite gifts from the ancient mountains and encouraged the fruitfulness of the everlasting hills. I am sure he planted the garden for his own family. The blessing reached across time to another.

When I was young, my family moved to the old farm in the Laurel Highlands of Pennsylvania. On our first visit, a rat ran across the the kitchen and into a hole in the wall. Plaster was falling off the interior walls and everything was dirty. Straggly, ugly bushes encircled the exterior. Vines climbing the exterior allowed minimal light through the windows.

Mom made the inside livable and then spent time in the yard. She found rose bushes under vine-covered, thorny, dead branches. She cut the disheveled bushes in the front yard to the ground, and dad tried, unsuccessfully, to dig them out. They pruned and trimmed as many other bushes and trees as they were able. I am sure we kids helped and hindered their efforts.

A few blooms surprised us our first summer there. In the second, we were amazed. The disheveled bushes in the front yard grew back as Annabelle Hydrangeas. They cascaded in heavy white blooms after being pruned. As the seasons passed, we realized that the person who planted the gardens planned the bloom times. All summer something bloomed somewhere on the property. Before the last blooms died off of one bush, color and fragrance burst forth from another.

Why the house was later neglected, I do not know. But hydrangeas are planted around my home today. In a neglected world with corners that still seem haunted, memories of my parent's garden tell me to draw hope from the best the sun can bring forth and the finest gifts the earth yet offers.

By comparison, I wonder what things were lost with Eden. God blessed a corner of the world solely for the purpose of meeting with and delighting the heart of Adam. Since man's fall, it is man who needs to make the effort to meet the Lord in a special place. A quiet time or sanctuary with God only happens when our hearts prompt us to make it a priority. This was as true in Biblical times as it is today.

The Lord said to Moses, ". . . Tell (my people) to bring me an offering. You are to receive the offering for me from each man whose heart prompts him to give. . . Then have them make a sanctuary for me, and I will dwell among them." Exodus 25:1b - 2, 8

With no shovels or modern tools, growing a garden would have been a lot of work after the fall. With a huge expense of time and effort, man dug plots of land and planted seeds for their families. Only kings and pharaohs grew gardens to display their wealth. A few hundred years ago wealthy aristocracy purposely grew grand, manicured gardens merely for their beauty. They didn't have to grow their own food. The purpose for their gardens was very different from that of the common man.

A garden's original purpose may have been a home in a lovely place that met all Adam's needs. But I think God placed us in a garden at the beginning of the world, because in His omniscience God knew that man would need to work for such a place to exist after the fall.

The land is a special thing, if you care for it, it will care for you.
-Eldest, by Christopher Paolini

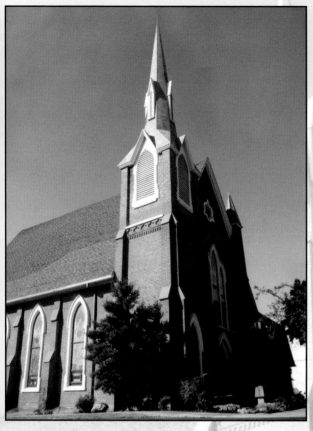

The earth is what it is without man's help. The ground covering most of the world is walked over, and passed by and barely noticed. In places like Alaska or New Zealand, it may be marveled over, but lies far beyond man's ability to govern. A garden needs a man. Though much of the rest of the world exists in a state of neglect, a garden only comes into being because someone has taken time to cultivate it. For this reason, a garden is an uncommon place in the world. Exceptional. It is the place we hang a banner of love, as in Song of Songs 2:3-5.

"Like an apple tree among the forest is my lover among young men. I delight to sit in his shade, and his fruit is sweet to my taste. He has taken me to the banquet hall, and his banner over me is love. Strengthen me with raisins; refresh me with apples, for I am faint with love." Songs of Songs 2:3-5

Gardened territories of life hold all we are willing to fight for and constitute places to care for it. They require consistent, intentional attention. They are places that need to be nurtured, tended and fussed over. They may not always be places we work, or places we live, but they are always places where someone feels at home. These areas of our lives don't have to be a source of income. They will always be priceless sources of joy. They are collections we celebrate, muse over and treasure. Though the original intention for Eden may have been a place we could spend our lives, gardens now are the places we spend each moment we can steal away from the world into which we have been cast. They are special, if only to the gardener and those who enter their boundaries. Every stone that borders such places and makes up their walls is cherished.

"For your people love every stone in her walls and cherish even the dust in her streets." Psalms 102:14

A garden is never an accident. It is purposely set apart and guarded by choice. It is a sanctuary, a corner of heaven formed on purpose. A gardener cultivates it to create something worth sharing.

"But my own vineyard is mine to give . . ." Song of Songs 8:12

Each ingredient forms what we see each time we enter a garden's boundaries. Soil and seed make up the smallest parts. Unnoticed laws of nature, such as climate and bloom time effect growth. Palm trees and spruce trees cannot survive in the same climate. Coneflowers simply will not bloom when daffodils do.

Different flowers and foliage lead the eye from one plant to another and around corners in a garden bed. Colors nest under canopies and cascade over arbors. Intimate hopes are aroused as sounds echo through fragrances suspended in the air. Isaac in Genesis experienced this even in his old age.

"So he (Isaac) went to him and kissed him. When Isaac caught the smell of his clothes, he blessed him and said, 'Ah, the smell of my son is like the smell of a field that the LORD has blessed. May God give you of heaven's dew and of the earth's richness-...'"
Genesis 27:27

The intrinsic details of each leaf create texture. The light the garden captures, scatters, or shades sets its mood. A mountainous backdrop humbles and intensifies splendor blanketing a path through one. Rings ripple and reflect its raptures in water.

"And these are but the outer fringe of His works; how faint the whisper we hear of Him! Who then can understand the thunder of His power?"
Job 26:14

ne is hardly aware of the imposing presence of tall trees and lush shrubs in any given oment. Evergreens offer windbreaks and shelter. Cut down or dug out, environment is ermanently changed. Voids echo their absence.

ees offer shade from the sun's heat and a covering from rain showers. How light ightens in the morning and casts shadows as the sun passes over them constantly nanges and recreates mood and atmosphere. Gardens are different every season of the ear. Elements all reminding us of God's ever-caring hand.

"...over all the glory will be a canopy. It will be a shelter and shade from the heat of the day and a refuge and hiding place from the storm and rain." Isaiah 4:5b-6

  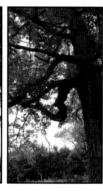

planned garden may include: places for retreat, water and flection, lines and shapes, symmetry and imbalance, texture nd shadow, depth and balance, and something that leads your e over the landscape and will not let it rest. The stillness of a ond is no less alluring than that of fountains and waterfalls. The oundance of flowers and foliage is no less important than that of egetables and fruits. All the elements of a garden come together eating a sanctuary from the world.

or some, especially men, a garden is a backyard to mow. I tease y husband about the men in our small community. If one of em is mowing when he gets home, my husband jumps on his actor as soon as dinner is over. It is not uncommon to see four ut of the six of them all mowing at the same time on the same ay. It seems like a competition, but I can't figure out how they etermine who wins. I make fun, but he reminds me that if he d not mow, no one would see my flowering beds. I think it is a

man's way of spending time alone in quiet beauty, circling their boundaries over and over. I am sure an ancient, unspoiled beauty is what draws us to state and national forests. Who can get enough of walking in God-given splendor untarnished by men? Yet, they only remain unchanged, because they are guarded.

• Was a place ever special because someone prepared and tended it for you?

• Has the transformation of a neglected property ever surprised you with uncovered secrets?

• Is your home a sanctuary that meets all you and your family's needs for relaxation and restoration? Is God celebrated in it?

• Have you ever caught the smell of a field the Lord has blessed?

• Do you have a favorite garden spot that is affected by the light?

• Has a tree ever fallen and changed a place you loved? What grew in its place?

Purpose...

Each garden has purpose. They are places gardeners needs. Even when I worked long days in other people's gardens, when I got home, I would often stay out until the sun set and work in my own. I love walking on my stone walks and weeding in my bare feet as the warmth of the sun radiates from the stones. I worship the Lord for flowers He left here for me and draw hope from their beauty. My garden always seems to be waiting for me. I am refreshed as much as the garden is by my time within it.

I love flowers that surprise my family and guests when they take notice. I have thought that when the year comes that I do not survive to spring; my flowers will embrace my family for me.

Gardens have great and simple purposes. They can be many things all at the same time and offer different gifts in different seasons. There are gardens tended to feed the body and gardens cultivated to feed the eye. There are knot gardens for balance and concentration, herb gardens for health, and water gardens for ponds. Rain gardens work in wet spots and rock gardens in dry hard places. They are places to rest by a fire's warmth, sanctuaries to dream beneath the stars or listen to a fountain. They are places to walk and places to work, places to be thoughtful and places to be useful.

To have a garden implies a relationship. Within a garden there is a constant battle meeting a place of tranquility. They are time consuming and demanding. Interaction and familiarity form one. Superficial love that talks and does not act allows them to disappear. They permit us to dig deep beneath the surface and branch out above the ground. As long as they are carefully tended, their nature is to give.

Gardens work very much like friendships. Some friendships feed my soul and others challenge my mind. I need friends to help me stay focused and others to take me away from routines. Some friends encourage me to be healthy and others encourage me to explore. I need every one of them to survive both drought and rainy seasons. They are worth working for and investing in. Hopefully, my own garden offers beauty and some form of harvest to those who call me friend.

In summer, I rarely go to a neighbor's home without flowers, but I never have any veggies to share. I am terrible at growing vegetables. I spend too much time in my flower beds. I know cucumbers, onions, radishes and carrots are good for me, even if I whine about eating them. I always loved the fresh jams, pies and canned sweet pickles my mom made from what we grew.

We are cautioned in Matthew to be careful what our lives produce. *Jesus said, "By their fruit you will recognize them." Matthew 7:20*

Some of us like things very simple. A daylily in a corner of a home or lilac planted on the outskirts of a lawn may be all the garden one desires as one friend is all we really need. Low-maintenance gardens have appeal. It is easy to forget the added weeding, watering and mulching that comes with each new bed. But there is more promise for beautiful things in and out of every season if work doesn't frighten you away. The balance between beauty and labor is tricky. Peace can be lost if you never have time to relax within too many garden beds.

My husband always buys me a tree or plant for Mother's Day. One year I asked him to buy benches and tables. Any more plants or mulch beds would take away the joy. Our gardens, as our friendships, should be places we can heal, celebrate, or find ourselves. They provide something we cannot find anywhere else in the world. At the same time we are cultivating them, our families are nurtured. Our roots set in deeply and anchor our hearts.

Precious and pleasant riches fill a garden when wisdom is its maker and understanding is its master.

"By wisdom a (garden) is built, and through understanding it is established; through knowledge its (corners) are filled with rare and beautiful treasures." Proverbs 24:3-4

• Do your relationships allow you to dig in and branch out? Challenge you and offer you a place of tranquility at the same time?

• What do you share most with your friend's flowers or fruit from your life's garden?

• Have you ever been discouraged by the work it takes to keep a garden (friendship) beautiful? Has a fear of work kept you from planting?

• Have you ever cultivated precious and pleasant riches when you planted seeds despite your fears?

Children...

"People were bringing little children to Jesus to have Him touch them, but the disciples rebuked them. When He saw this, He was indignant. He said to them, 'Let the little children come to me, and do not hinder them, for the kingdom of God belongs to such as these. I tell you the truth, anyone who will not receive the kingdom of God like a little child will never enter it.'" Mark 10:14b-15

I have learned most of what I know about a garden's care from Susan Murray. She is a gardener, an incredibly talented horticulturist and artist. She is my mentor and friend of many years. Her passion is evident in all she touches. The underlying focus in her designs is the enjoyment of children who will visit each garden. She often says that if a garden is not planted for children, it misses its purpose.

Gardeners can at times be overprotective of their garden treasures, but flowers, fireflies, and fragrance summon children. Every garden on earth should freely invite children. Childhood needs wonder, beauty and magic. Too often we protect flowers from harm and forget to hearten and inspire children within blossoms. Every child should have the chance to play along the paths and around the flowers and shrubs of some enchanted place, the kind of place they go to bed dreaming about and are never able to forget. It will never matter that they can't remember where it was a week later. That is the magic gardens offer each spring and after every destructive storm. Even if it never grows back (even if we never revisit it) its memory will always be there. Awareness that Eden ever existed testifies to a garden's permanence. I think that is the point. It is unforgettable, because we are not supposed to let its glory slip from our hearts.

"Only be careful, and watch yourselves closely so that you do not forget the things your eyes have seen or let them slip from your heart as long as you live. Teach them to your children and to their children after them." Deuteronomy 4:9

"He is the maker of Bear and Orion, the Pleiades and the constellations of the south. He performs miracles that cannot be counted." Job 9:9-11

When all the rest of the world takes our innocence away, gardens recreate it. Unmistakable beauty so pure and simple is unforgettable in its rarity. Like few other things, a garden can restore childlike wonder. When all the rest of the world would swallow us up, frustrate, and change us, beautiful places remind us where we came from. They communicate what we are, why we were created, and show us exactly who we are when we can't remember it any more. The things that are best at doing this all have a connection to God's gifts. He is the maker of big, fluffy snowflakes, cloud formations, rainbows, fireflies, sunrises, sunsets, stars, ocean waves and coral reefs. They all testify to God's signature over earth and above it. Perhaps that is why we are mesmerized by campfires, they are fueled by growing rings that record years gifted the world. His wonders never cease.

• Was there a garden place in your childhood that you have favorite memories of? Do you wish there were?

• Are there things that you almost forgot and had trouble sharing from your heart with young people in your life?

• Have you ever found restoration in a quiet garden sanctuary?

Growth...

Even with careful attention, a garden is always changing. And what defines a garden changes as a person grows. Children, because of size and age, will see and understand a garden from a different perspective than an adult. My mom's favorite memory of her grandma's garden is looking up into the blossoms of a foxglove. That is not a view adults see except on their knees. Childhood sees flowers, maybe colors and possibly fruit on vines. A child may not even notice the garden as a whole. Growing up in cities and towns, entering a landscape of wonder has tremendous power to wake what lies dormant.

It is a surprising and memorable, as well as valuable experience, to be lost in the woods any time. Not till we are lost, in other words, not till we have lost the world, do we begin to find ourselves and realize where we are..."
– Henry David Thoreau

I worked in gardens for a while before I thought of things other than flowers that made an impact in one. After a while design was advantageous. Structure, repetition and order became important and architectural elements became highlights. Learning about form and textures in oriental gardens opened new perceptions. A garden has incalculable, unnoticed elements that make a huge difference. It has elements that can seem timeless. Flowers and colors are only visitors. There are tiny seedlings, mosses, lichens that I have come to love. Most I hardly noticed until a trained eye taught me to take note of them and learn their names.

There are things to love about gardens, but with those come numerous battles that never end. Flowers need to be deadheaded as they expire. Herbs can be rambunctious. Seeds battle the drought. The water wrestles with sun. Yet the shade tries to extinguish the light. Winter even contradicts the summer. Yet each element is essential and immeasurable in importance for what it brings to the garden and can teach a man. Yes, even weeds have something to offer. In a perfect world, I think, rabbits would have fed on them alone, and weed seeds would have fed birds. All would have been satisfied, but God's designs are not limited by my imagination. God has a reason for everything. His word tells us the animals, fish or birds do not dispute this truth.

ardens reveal completely different things to people in different ages of life. Even the absence of its beauty or the chaos of the ilderness can help see clearly. Desperation focuses on very ifferent things than prosperity. In youth, a person will draw one erspective. After raising a family, they can come away with an ntirely different compilation of impressions. Men and women e drawn to different elements. Gather all parties together in ne place to talk about what each sees and each will add to the ther's experience.

"But ask the animals, and they will teach you, or the birds of the air, and they will tell you; or speak to the earth, and it will teach you, or let the fish of the sea inform you. Which of these does not know that the hand of the Lord has done this? In His hand is the life of every creature and the breath of all mankind."
Job 12:7-10

As things grow, they change. It is not always easy to anticipate how age and time will change a garden. My husband says I move my plants around like furniture. He is wrong. I move my furnishings more often. I like change. Change is often a good thing. Most of my plants have been transplanted at least once. Even with my experience and plant labels, sometimes I plant things too close or in the wrong place.

In life I make the same mistakes. Transplanting can be hard, but to flourish and reach our potential sometimes it is necessary in life as in a garden. Some shadows shelter us and others stifle us. Growth changes us. There will be things we overlook in youth that we will want in adult life. Order and structure will be helpful at times, but we will constantly long for the surprise of the next thing to bloom. There will be things we overlook and then learn to appreciate. Growth is learning with each change.

• Has your definition of beauty changed as you have aged? Has your taste in fruit or perspective of a garden been altered by another's point of view?

• Which garden battle best parallels your life today? Do you long for light, shade, water, more room or climate change? Battle weeds or chewing rodents?

• Do you like change? Is it possible that transplanting would help or hinder your growth or that of someone you are concerned for?

Roots...

A root is understood to be part of a plant that grows underground. To root up is to dig out completely, yet to take root is to grow or become fixed. Root hairs are tiny, yet absorb most of the water and minerals even the tallest of trees need. Roots are unseen connections between visible parts of a plant and its environment. They are marvelous, little-understood structures for anchorage, support and the gathering of nutrients. If roots are not healthy outcropping are not either.

Deep and broad roots hold the soil in place. Tap roots grow deep for anchorage, but they can fail to gather enough moisture in time of drought. Many plants have broad branching roots that cover just a few inches below the surface. They gather the most moisture, but are easily torn up and suffocated by foot traffic.

The environment in which a plant grows can affect which type of root it develops. For example, a tree deep in the forest has shallower and weaker roots than a tree that grows alone on a windy plane. The forest protects woodland trees from strong winds. However, because it is competing for water, its roots are wide ranging to gather water quickly. This is necessary for survival, but these roots are shallower and less secure. The wind rocking a tree back and forth makes it stronger and its roots anchor deeply. There is no competition for water but also no shelter from sunlight. White river birch grow best in damp places in thick stands with other birch. They will never be long-lived trees planted individually. They depend on the roots of other birch roots around them to be strong.

Our lives are formed in places of isolation and crowded assemblies. Both conditions affect our foundations. Adversity can be good for us. It sends our roots deep into foundations that sustain us. Competing for time in the spotlight and refreshment in overcrowded arenas can be detrimental to growth. Because they are hidden, it's not always obvious whether they anchor bitterness or strength.

Choosing a place to put down roots is one of the most important choices in life's story. The best soil may not always appear healthy and rock free. In Genesis, Lot's choice for the better land was not beneficial for him.

"So Lot looked up and saw that the whole plane of Jordan was well watered, like the garden of the Lord . . . (This is before the Lord destroyed Sodom and Gomorrah.) So Lot chose for himself the whole plain of the Jordan..." Genesis 13:10-11a

The choices available for a foundation can be overwhelming in this fast-paced, mobile age. Yet I think God cares very much where we put down roots. Often people comment that it is ridiculous to imagine Israelites wandering for 40 years in a desert. Yet, even today, many spend lifetimes wandering, trying to find the Promised Land. Some wander far longer than 40 years. For far too many, it is never found. Without The Creator, we all die searching.

In The Four Loves, C. S. Lewis says, The highest does not stand without the lowest. A plant must have roots below as well as sunlight above and the roots must be grubby. Roots may be unobservable. However, every visible part of a plant is dependent upon them. No plant could survive freezing temperatures or oppressive heat without their unseen power. Yet their job is grimy, dirty and unnoticed.

Mucky, soiled and disguised roots matter! Foundations should be

laid in places where we are willing to dig deep and get dirty developing. When passion is fueled by hidden passion being grimy while building it doesn't intimidate us, our outcropping will not dwindle in adverse conditions. Much like a plant only a fragment of a man is visible. Isaiah says this about Jesus:

"He grew up like a tender shoot, and like a root out of dry ground. He had no beauty or majesty to attract us to Him, nothing in H. appearance that we should desire Him." Isaiah 53:2

What held Him up? From what depths did He draw His invisible beauty? Man's survival depends upon a struggle against inward depths, while reaching for the Light of the World. As seed that falls on rocky soil, we dry up without grubby anchors.

"But since they have no root, they last only a short time. When trouble or persecution comes because of the word, they quickly fall away." Matthew 13:21

We all need gardens capable of anchoring dreams, hopes and futures.

"But blessed is the man who trusts in the Lord, whose confidence is in him. He will be like a tree planted by the water. That sends out its roots by the stream. It does not fear when heat comes; its leaves are always green. It has no worries in a year of drought and never fails to bear fruit."
Jeremiah 17:7-8

Our Creator chose a place and a perfect time in history for Christ. He has a place to plant each of us. Our undertaking is to anchor our hearts to His, soak up every drop of His love and reach for the Light of the World.

• What unseen connections to the world anchor your heart? Nourishes your soul?

• What parts of your root system were developed in isolation? What benefit did you get from roots grown in an assembly of friends?

• On what invisible anchor can we depend when the world storms around us? What unseen anchor helps us survive in seasons of drought or bitter cold?

Treasure in a Field. . .

"The kingdom of God is like hidden treasure in a field. When a man found it, he hid it again, and then in his joy went and sold all he had and bought that field." Matthew 13:44

When I taught this Sunday school lesson, the kids always wanted to know why the man hid the treasure again. I struggled with that myself. Most men would claim ownership and walk off with it if they could. But perhaps the treasure is permanent and fixed, living and overflowing. Isn't that the kind of treasure a person would sacrifice all to possess? It might sound crazy, but I think the treasure the man found in the field was a big muddy spot.

In the Middle East, the land changes drastically from one place to another. A desert, lush vineyard and rocky caves can all coexist within a small perimeter. In a time of drought, a muddy spot in the middle of a dusty, barren field would have been a sign of hope. It would have been a source of water. Since Biblical times, wells have been a foundation for all development.

I believe God put a longing in each man for a place with hidden treasure, because He has such a place for us. To find it, we must be searching for Him.

"Whom have I in heaven besides you? And earth has nothing I desire besides you." Psalms 73:25

The thing about treasure is we are all looking for it. Looking in the right place for it and identifying it when we unearth can be a bit tricky though. The Bible promises floodgates holding back heaven's riches, if we would but recognize where blessings find their source.

". . . Bring the whole tithe into the storehouse, that there may be food in my house. Test me in this," says the Lord Almighty, "and see if I will not throw open the floodgates of heaven and pour out so much blessing that you will not have room enough for it . . ." Malachi 3:8b-10

Treasure does not always look like we think it will. Since the fall, in truth, we generally misidentify it, and everything we hold back from God keeps us from it. The field we cultivate may be different, but the treasure is always the same. What makes it precious and rare is that it is not stagnant or unproductive. It does not fit in our pocket and cannot be carried over our shoulders. It cannot be restrained. It is a place we cultivate beauty and produce fruit. Eugene Patterson says, "It is no defect of a fountain that it is prone to overflow." It is a fountain that never dries up.

Treasure in Matthew 13 is connected to God. The place had value because God's blessing hovered was over it. It could not be taken from it. Because the place contained all that was cherished, an atmosphere of peace encompassed it.

We can only assume that it was hidden so that when the land was purchased, it could be bought at the value the present land owner placed upon it. The one who sought the treasure could never have purchased something he knew to be priceless though he spent a thousand lifetimes trying.

I am not sure if the man selling assigned too little value to the land or if he did not value the treasure it held. It may never have had value for him, because he was in the wrong place. It could not have been a place God wanted the seller to keep, because he was willing to let it go. Matthew 21 tells us God is not about to let His territories be wasted or withhold treasure they offer. He gives them to those searching and sacrificing for them.

"Therefore, I tell you that the kingdom of God will be taken away from you and given to a people who will produce its fruit."
Matthew 21:43

In Genesis 27, Jacob recognized a blessing that Esau took for granted. What his brother was willing to sell for a meal (Genesis 25:34) Jacob would not let go even when he had to wrestle God Himself and be wounded for it (Genesis 32:24-26). True treasure is worth being wounded for if necessary.

After we have given all we have for a place to build, the work has only just begun. Digging a well is hard work, and living things will depend upon water from it day after day. Such is the kingdom of heaven. It is not something we possess, it is something we draw from and share. It is a land with a well from which we draw refreshment and a river of hope. It is a place we can plant beauty and because of it we have the power to bear fruit.

God will ask us to give up every obstacle that keeps us from this type of blessing, because He loves us. The things we need to let go of for that to happen always make room for what waits behind the floodgates of heaven. It is only found in a place with room for God. We can't move God or His will for our life. We can claim it. We can let go of all the holds us back and begin where He waits for us.

" . . . If your presence does not go with us, do not send us up from here. How will anyone know that you are pleased with me and with your people unless you go with us? What will distinguish me and your people from all the other people on the face of the earth?" Exodus 33 : 15-16

If God is not in the gardens we are cultivating, we are putting roots down in the wrong place.

"But seek first His kingdom and His righteousness, and all these things will be given to you . . ."
Matthew 6 : 33

Have you ever encountered a metaphorical muddy spot that gave you hope?

Have you ever unearthed a treasure that you never imagined would become so precious to you?

What overflowing fountain of treasure would you sell everything for?

Have you ever been lead to a place that appeared lush, but felt foul? A place that appeared barren et blessings emanated from it?

Have you ever let go of something you hadn't recognized the value of? Fought God for blessing?

Share your garden with a child, be it a relative or neighborhood kid or take a child to a garden. Share our memories. They will make an impact.

GARDEN
Always

Concerning Gardening...

My home church is full of small circles some with couples, some men and others with women. My closest friendships are rooted in them. I am weeded and fertilized through them. Much of what I have learned about the Lord and myself, I gain from them. Community groups surround many of us with support and encouragement. A garden circle is one such group that cultivated my heart.

Our church existed in a small historical building for more than 100 years. We have since grown into a new location. For quite a few years its historic location had very simple garden beds, but weeds peppered them with indifference. There was so much myrtle that nothing else grew. Rhododendron sticks with a few leaves dwindled. There was an untidy half-dead holly tree. The bushes were unkempt and a cement planter was filled with mossy soil. Our pastor asked for help.

A circle of friends formed a garden circle and pulled together a design. We calculated the details of clean up, hauling and expenses. A work date was put in the bulletin. We asked for tools, pickups, and willing hearts to join us.

We joined hands and bowed our heads before working. We cleaned up and planted in one day. Parents brought their kids. Their willing hearts did more than anyone expected. I wonder what seeds they collected and stored as memories in those little hearts. Many, who were physically unable to work, brought food and refreshments. Then mulch covered the working wounds at the end of the day. The Lord's people put down roots and bloomed. The church was transformed and more beautiful that year than any since.

A Garden Circle has surrounded our congregation since. We gather names of elderly, sick, homebound, wounded, grieving and even new moms who need help. One very special day we spent with a mother whose child we knew would be lost. A birth defect only allowed her to live a few precious hours after birth. The garden we planted still reminds us all of a special day with her unborn child. We still gather for visits and bring flowers and food. I hope long after this group dissolves, a new garden circle will still function as a ministry. Being a part of it has been one of the riches rewards of my life.

It is hard to stir up excitement about each project, so I send out emails and ask everyone to dust off their wings and tell them not to forget their sparkle. We have become garden fairies, encouraging God's kingdom. The number of women who circle homes of those in need with wings fluttering and a luminescent glow is amazing. Men come, but thus far they have chosen to remain men. It has been mentioned, men, may perhaps, at times, resemble, ogres more closely than fairies, but only when they disagree with us or grumble about our ideas. We play and tease each other, but ultimately, plant love, add fertilizer and water struggling souls in need of personal contact and a sincere embrace.

An extraordinary love was seen in the transformation of deserted and neglected gardens. Then the people of the Lord will know that they are like those Isaiah 62:12 called,

"Sought After, the city no longer deserted."

From this garden circle, energy and encouragement flows. I will not say that planning is not hard and sometimes mistakes are made. But the lonely are not forgotten and the wounded are blessed.

Genesis assures us that we are all called to cultivate and keep whatever territory the Lord gives us. "The Lord took the man and put him into the Garden of Eden to cultivate it and keep it." Genesis 2:15 NASB Everyone gardens in different ways and for different reasons, but our purpose for it is the same. Through the loving-kindness of our hands He cares for the work of His.

"The Lord will accomplish what concerns me; Your loving-kindness, O Lord, is everlasting; Do not forsake the works of Your hands."
Psalms 138:8

• Do you have a circle of friends that supports each other and reaches out to embrace others?

• Is there any one component of your life that gives you wings and a luminescent that others can see?

Abad and Shamar . . .

"The Lord took the man and put him into the Garden of Eden to cultivate it (abad) and keep it (shamar)." Genesis 2:15

In Genesis 2:15, the Hebrew word for the Engli[sh] cultivate is abad and for keep it is shamar. The[y] are tied together. The two Hebrew words mea[n] more than our simple words communicate. Their meanings together encompass keeping, guarding, and protecting. They thread togethe[r] the idea of the observation, preservation and restraint. They include the idea of treasuring u[p] as in memory, and the celebration of the eart[h's] gifts.

In large public gardens everything looks so beautiful and easy. But gardens are not easy. Work you spend all week doing is constantly overrun. The flower's glory one month points t[o] deadheading the next. One downpour or win[d] storm can wash away topsoil and smash long-established plants. All of it collapses under the weight of winter here in Pennsylvania. I imagi[ne] in places like Texas, it all crumbles or goes dormant in times of severe heat.

It is one thing to admire a garden's beauty; quite another to take on labor pains required [to] birth our own. It does not turn 18 and move o[ut]. A garden's survival depends upon groundwor[k] and commitment.

Genesis tells us man's labor is as important to the earth's care, as rain is for its growth.

" . . . and no shrub of the field had yet appeared on the earth and no plant of the field had yet sprung up, for the Lord God had not yet sent rain on the earth and there was no man to work the ground . . ." Genesis 2:5-7

Gardens require seeds, water and light, but only survive with abad and shamar. Gardening is paramount, not to the earth's survival, but to its enjoyment. It's well being and fruitfulness are connected to the Lord's blessing as much as our willingness to fight for it. Henry Van Dyke said,

"There is a loftier ambition than merely to stand high in the world, it is to stoop down and lift mankind a little higher."

In very much the same way we live, we keep a garden. Gardening is not accepting things the way they are. Rather, it is encouraging the best of things that are. It is inviting the pieces of the world we love into a small space: things that fascinate the eyes, fingertips, nose, taste buds and appetites. Gardening is an embrace of all we keep. It is actively being thankful for life and treasuring each of its gifts.

• Have you ever been surprised by how much work a garden demands? Been fascinated by all it gives?

• Do you mind all the time you spend stooped over in your garden? Do you find fragments of joy fussing over your life's garden?

• What do you treasure and celebrate about the world God gave you?

Cultivating and Keeping...

The first years of our marriage fill the halls and rooms of one apartment after another. We build one home from the ground up, hoping to stay forever, but God closed the gate of that home as well. Packing up brought sorrows but redecorating each house hopeful joys. When the kids were little any place we were all together was home. As they grew, the dynamics of each transition became scarier. I knew that each conversion was a time of plowing, and so I tried to keep my heart open to the lessons and to people God placed in our path along the way.

"Does he not plant wheat in its place, barley in its plot, and spelt in its field? His God instructs him and teaches him the right way." Isaiah 28:25b-26

Every time I could, I dug up plants and brought them with us. Those balls of roots and injured foliage packed in buckets of soil helped remind me that a piece of each home and a part of its memories would survived. So in every location I planted new gardens or restored the ones that had been left. I encouraged the plants to stretch their roots in new beds. It has been encouraging to drive past some of those places and see that they are still cared for and very sad when they are not.

This is what gardening is. It is taking charge of the places under our care and trellising memories to them. A big part of being a gardener is letting go when it is necessary. It is kneeling and starting again every time we have to. Tangible beauty falls dormant each fall and continues to regrow each spring since Eden. More times than can be counted blooms come in the moments they are most needed. The same seasons bring the same weather patterns, smells and memories along. There are times this is comforting and times it can break a heart all over again. But with each break it brings healing and new growth wraps itself around what was broken. Year after year life has a way of intertwining new memories with old, trellising the moments of life in lush vegetation and blooms of hope despite what has been stilled.

Jeremiah tells us clearly what the world would be like if no one cared.

"It will be made a wasteland, parched and desolate before me; the whole land will be laid waste because there is no one who cares . . ." Jeremiah 12:11b

When no one cultivates, thorns are creeping. Keeping a garden is akin to taking out the garbage, changing diapers, or caring for a constantly lonely soul. All the work can wear a person down.

"They will sow wheat but reap thorns; they will wear themselves out but gain nothing." Jeremiah 12:13a

Like life, gardening can get monotonous and boring. It is some balance of cooperation and irritation. It is not a novel or movie; with a quick solution and clean ending. But the drudgery comes with a purpose. Each act points toward a belief in something better. This is especially true when caring for others. When people care for each other roots go so much deeper and what can grow that much bigger. When communities or families take up a passion and all work together, branches extend further and fruit is more abundant.

Cultivation is fixing and tidying one day and returning to do it over again the next. Its importance can be very hard to see when it is well done. Turmoil and joy can get entangled in it and be hard to separate. Solomon speaks to this truth in multiple places in Ecclesiastes.

". . . I wanted to see what was worthwhile for men to do under the heaven during the years of their lives. . . I undertook great projects: I built houses for myself and planted vineyards. I made gardens and parks and planted all kinds of fruit trees in them. I made reservoirs to water groves of flourishing trees . . ." Ecclesiastes 2:3b-6

". . . My heart took delight in all of my work, and this was the reward for all my labor. Yet when I surveyed all that my hands had done and what I had toiled to achieve, everything was meaningless, a chasing after the wind; nothing was gained under the sun." Ecclesiastes 2:10b-11

How do we care for and keep a garden? Even as Adam in a perfect garden did, turn one shovel of soil and then another. Pull one weed and then another. Plant one seed and then another. Water one day and return to do so the next. And when it won't stop raining, gardening is the willingness to change tactics. Either we divert water or install a fountain. Why we choose to garden is rooted in our purpose. It is to make the world a better place. Once a gardener is born she can't go back to being a desert or a weeded jungle without slowly dying herself.

I listen to RZIM Ministries all the time. Ravi Zacharias says, "We were made for worship and that misdirected worship is one of life's great tragedies." No created thing can bear the weight of a man's deepest desire, so we need to be sure that what we are worshipping deserves the ecstasy and heartbreak it is sure to deliver. Pleasure leaves us empty when it has delivered what it can. Gardening is an act of worship that, "Brings us to our knees and lifts you up to the stars." Tim Keller.

The places we work, call home, and worship become havens from the world when our hearts are open to the lessons God has for the gardener in each of us. We wear ourselves out unless our desire is to keep what God has given us and cultivate the purpose of those God has planted alongside us.

We live in deeds not years; in thoughts, not breaths;
In feelings, not figures on a dial. We should count time by heart throbs.
He most lives who thinks most, feels the noblest, acts the best."
Philip James Baile

• Have you ever had to walk away from something you spent years developing? Or watched it crumble? How hard was it to start over?

• Can drudgery have a purpose? Why should it be shared?

• How do we find meaning in deeds, thoughts, breaths, feelings?

Naming...

"A good name is more desirable than great riches; to be esteemed is better than silver or gold." Proverbs 22:1

In Genesis 2:19, Adam chose names for the animals. God was deliberate about this assignment. I don't know when plants were first identified by traits, but I do know successful gardeners know plant names. Naming things is an important initial step. Nomenclature delineates relationships and reputations. It is part of taking ownership. A name easily passes on information to others who will need understanding.

Names define strengths and harmful characteristics up front. A plant's name tells what soil type it grows in, what watering needs and light requirements it cannot survive without. With a name a gardener can take steps to plant in a place with room for all the potential imprinted in the genetic makeup. For example wisteria plants, trumpet vines, or weeping willows are beautiful. But we must be careful where we plant them. Each of these plants can be beautiful in the right place. Each of these planted incorrectly and without knowledge of its reputation, can be very destructive. Willow roots crush septic lines. Trumpet vine roots crack foundations and wisteria vines crush anything weaker than rebar when improperly encouraged to climb. The same rules apply to a garden and life.

Names give us dominion and understanding. Entering into an interaction with a named entity, we know the good, bad and little understood characteristics. A name tells us what benefits or hinders growth in horticulture. In the same way a person's name implies what loves are attached to him and what hurts have formed her. With this simple word we best know how to bless and avoid hurting. For this reason, the Lord calls each of us by our own names.

"... I have called you by name, you are mine. ..." Isaiah 43:1b NLT

The importance of a name was no less significant to in Old Testament times as today. Job longs to be remembered by the reputation his name once held in his village.

"They waited for me as for showers and drank in my words as the spring rain. When I smiled at them, they scarcely believed it; the light of my face was precious to them." Job 29 : 23 -24

I loved a big tree on our property when I was growing up. The air under it was cool in summer, and I often climbed into it, savoring the light in its leaves and a quiet escape. Beneath its branches my dreams were as limitless as fall foliage floating down from its canopy. In college horticulture classes, I identified the tree as a beech. I felt like I was being introduced to it for the first time, though I had known it my whole life.

A few years ago we needed a new mechanic. I took a chance on a place I wasn't sure about. When I brought my car in for the first time, the owner came out and asked me if I was my mother's daughter. He had known my grandfather and said that he loved and missed him. He still thought of him every day. My grandfather had died two years earlier. He told me a couple of funny stories and took really good care of my car. I could not stop smiling as I left. That man's memory of my family's name, assured me that I would never need another mechanic.

Much like family, trees hold prominent places in the landscape of our lives. Knowing their names empowers us to discover their secrets. I think that is why I enjoyed putting together a leaf project with my kids in junior high. We explored Somerset County and searched for the well kept secret names of trees they had walked beneath since they were little.

Driving around looking for different trees and gathering leaves was the easy part. Identifying their unfamiliar names was sometimes quite hard. There were a few frustrating moments. We mixed up leaves, damaged them and had to find others. But we learned a lot about the hardwood of southwestern Pennsylvania as we explored. From an early age my kids were introduced to a new way of looking at their environment, something I wish I had discovered sooner.

Much like a person's surname, a tree's name reveals things about its history, roots and heartwood. Oaks are all strong and solid, but slow growing. White birch grow well in wet spots, and live longer in groupings than as single specimens. Discovering that sassafras roots are used to make root beer is just plain cool, but they also have three differently shaped leaves, two that look like mittens. I remember tasting birch beer from a bottle long before I discovered the flavor from a soft wood that grows locally. Maple syrup is worth celebrating because of our beloved maple trees.

If you think about it, you can probably remember much of your life in trees. I remember being afraid of the creepy-fall seed pods of a catalpa that grew over my childhood home, but I loved its fragrant flowers. I remember mom making cherry pies from the sour cherries that grew in our backyard, playing under apple trees in a very old orchard with my brothers, and making wreaths from spring blossoms with my sisters. I first saw a palm tree in Florida while visiting my grandparents for my sweet 16. I was married under a hawthorn tree, and my first home was shaded by a silver maple. Since then I have seen grand sycamores in New York City, lodgepole pines and aspen in Yellowstone, and I hope someday to smell the cedar forests of Lebanon as I walk beneath them.

Home to me now is a tree line of hickory, maple and oak trees. The native service berries outside my front windows are where my heart is also. They flower in the spring and invite cedar wax wings into my yard in the fall.

If you don't have a favorite tree it is simply because you haven't heard its name. Even gnarled old trees in the forest have names and stories to tell, if only imaginary tales about trolls and hidden treasures. And one of life's great treasures are stories told to teach and inspire wonder with children while we wandered the wild. Thinking back on that leaf project now, I know my kids learned a lot, but I was reminded of how much I had forgotten. Names make a difference. Too easily they dissipate like beloved fall leaves if their stories are not retold. Gathering leaves is a tangible way to hold life's special moments in our hands, and remember the trees that held us in their arms when we are small.

Names absorbs into them the story of our long history and whose we are now. Of course, they carry good and bad connotations. It is a good thing they do. They help us recognize known fears, joys and interests and tell us whether a relationship will be easy or hard. They warn us how to care for some people and avoid others. What hurts and what heals is revealed.

The Bible tells us God knows us each by name. Thus, He knows us each by everything that makes us uniquely ourselves. He cares for us throughout our lives according to exactly what we need to thrive and fruit. The God who gave each unique characteristic has not forgotten a single name or priceless thing about any one of us. Gardening is giving names to what we treasure.

"I have chosen and consecrated this temple so that my Name may be there forever. My eyes and my heart will always be there." 2 Chronicles: 7:16

• In your day to day choices do you worry about your good name?

• Do names ever hide anything?

• Have family lineages or stories ever been important to you? Why? When aren't they?

• Can you think of a time you did something for someone you'd never met, because you knew their name?

Seasons Remain...

"Flowers appear on the earth; the season of singing has come...
The fig tree forms its early fruit the blossoming vines spread their fragrance..."
Song of Songs 2:12

Like many married couples who married young, my husband of 20 - some years and I grew together. Sometimes we grew close. Sometimes we grew apart. The seasons of life changed the environment around us while sculpting us into individuals and crafting our relationship.

There were seasons along the way that were not long enough and others we thought would never end. I waited for him at times and he stayed a few times when it may have been easier to leave. Believing in more than the present circumstances was necessary. To linger anywhere or with any soul will promise continual change in person and place. Seasons of both blessing and trial will follow one and another. Neither my husband nor I remain the person we were when we first married.

Relationships go through seasons much like gardens. Times of excitement when, new life springs up and everything is new, can be like the spring. But a summer season with either growth or drought is possible following that. Growth leads to pruning, division and transplant in the long run. Flowers need deadheading. For example, we had our son in the first year of our marriage. It was a time of huge transitions for me. It took me a while to put down roots and longer to bloom again.

Sometimes everything changes and transition is unstoppable. Blooms can disappear in one harsh frost or last for weeks under the right conditions. A relationship that lasts makes choices to stay despite change. Gardens take time. They grow over multiple seasons. There is an old saying about gardens I like to remember. "The first year they sleep, the second year they creep the third year they leap." Seasons of transition follow times of stillness and waiting. Dormancy is just as important as growth at times. There is often a lot more going on beneath the surface than can be understood. Because my husband loved me through my time of slow growth, it was easier to bloom and fruit for him once new roots took hold. He remained season in and season out. He still rakes leaves in the fall with me, and I wait out winter beside him.

We've survived our own personal hurricane and perhaps a tornado or two. Then bent down and started to rebuild. Some of what was broken we were able to repair. There were some ruins that new vines have grown over and covered from view. They remain a part of the garden and add their own beautiful difference to our lives.

Life changes all the time. One moment life can be rich with color and the next naked, wind-blown branches can shiver in muddy gray. When it is hot and dry, day to day survival is all we hope for and droughts usually come in seasons we expected growth. Watching what we've worked for dissolve into a bed of dust can wither something within us. Winter sets into our hearts in environments of grief. It is often hard to hope for change as our world becomes as cold as frozen ground. But times like this make spring's sweet warmth more precious. They are a part of the blessing.

"See! The winter is past; the rains are over and gone."
Song of Songs 2:11

Spring flowers are noticed by more people and more needed perhaps after the long winter. Flowers in the summer are abundant and sometimes how pretty they are can be overlooked, because there are so many. They lose their wonder. The beauty of each individu is lost.

It is not the same thing to say, "I love the garde I have seen at such and such a place," and to s "I love the gardens that surround my home." Unforgettable as they may be, visited gardens do not cultivate us. Gardens our own hands have tended and our hearts have dreamed send roots down and cover seeds deep withi us. Seasons of preservation required guarded effort.

Seasons of growth require pruning and training. Seasons of harvest bring rich blessing, but it's possible to meet sorrow in them. Crops are cut. Fields get mowed down, and the landscape changes. Fruit falls fro trees and what is not collected rots. Focusing on good things in these times, as the lover in Song of Songs focused on the henna blossoms in vineyard of En Gedi, is paramount. Song of Songs 1:14 En Gedi was known as an oasis in a harsh environment. Small glimpses of beauty stood out because of stark conditions that surrounded them.

Gardening allows us to enjoy beauty when it stands before us and to believe in it when it is hidden from our sight. When the world is fresh and green it is easy to believe it will always be so. But the world chang seasons come and go. I love peonies, but they only bloom for a short period each summer. Then for more than 11 months I have to wait for them to open their petals again. A long wait is the only promise for another year's blossoms. If the only thing I loved about my garden were the blooms, it would disappoint me. If however, peony greens als served as a backdrop to ground covers and a foreground to shrubs or trees, my love would branch that much wider for them and things they highlighted. If shapes and figures within my garden blanketed in snov also gave me hope, a garden's gifts would reach that much deeper.

There is no exact way to garden. Every garden is affected by soil makeup, zoning and seasons differently and life plays out on a different dimension. There is no rule book for remaining in a place or with someone we have committed to. Letting go one day may be necessary, but nothing can take the memories we keep of the part of our self that rooted and bloomed in them. It is easy to envy other people's marriages, but ultimately marriage is a promise to love each other while we are forming, struggling and blooming. Seasons will change it.

Gardening empowers us to change our world for the better. My thoughts and prayer are poured out before the Lord in the stillness of my garden. As springtime is sweeter after surviving winter, struggles bring blessings found in no other way. There is no promise we will always be able to show love or that it will always be received. Recognizing what the Creator has given and remaining faithful despite seasons of rejection and emptiness is the hope for good things in the future.

- Can you describe a time when someone who gave you time to put down roots? Has anyone ever needed time from you? Were you glad you waited?

- What is worth waiting for in your garden? What fruit or blooms are you waiting for in life?

- Why do some people forget to savor the diverse elements found in a garden as well as from our lives and relationships?

- How important is it to remember summer in the midst of winter? To believe in beauty when it is dormant?

- Can you think of a time when you lingered somewhere and found the wait was worth it?

Guarding, Protecting, Preserving...

(The Kingdom of Heaven) "is like a mustard seed, which a man took and planted in his garden. It grew and became a tree, and the birds perched in its branches."
Luke 13:19

Beautiful places can harbor monsters. Those monsters are not always ugly. Satan didn't tempt Eve to be bad. He tempted her to be like God. Beauty is a battleground for God and Satan. The beauty in a garden, its lush leaves, and colorful blooms can hide snakes much the same as Eden was haunted by Satan. They slither across the ground and hid in the branches.

"Now the serpent was more crafty than any of the other animals the Lord had made . . .
Genesis 3:1a

The most constant battle of life is not with another person or even the world. It is a battle with the creature walking around in our own flesh and sharing one heart. "But if your eyes are bad, your whole body will be full of darkness. If then the light within you is darkness, how great is that darkness? Matthew 6:23 Our hearts, like a garden, grow and change. The darkness is twisted by neglect and havoc.

"What is twisted cannot be straightened; what is lacking cannot be counted."
Ecclesiastes 1:15

Life has nooks and crannies, branches and ditches. The light of Christ overcomes the darkness, and counts what is lacking.

"I have told you these things, so that in me you may have peace. In this world you will have trouble. But take heart! I have overcome the world." John 6:33

The world is not fair. Perfect days are few. Gardening some days is hot and sweaty. Other days it's cold or a battle against wind. There are buggy days, weeding days, mulching days, and digging days. But a garden is worth each skirmish. A life well spent isn't perfect all the time, but it is full of special moments and special places.

Gardening is the willingness to suffer and acts kindly even confronting harsh realities. 1 Corinthians 13:4a NKJV I love summer mornings, looking out my window to see my husband watering the garden. Before heat comes, he pours himself out for our marriage in the same way. He showers affection even when I am grouchy. He knows that gentleness and patience grow more than force and will.

Gardening is working toward the survival of small saplings and grand old oak trees. To guard the garden is to stand sentry over it, as by a gate. It is building a wall or defining a border to protect what is within. It can be as gentle as a hen's wings over her brood or covering tomato plants and annuals with our sweaters and blankets to protect them from the frost. It is weeding all day and planting as many seeds as the ground can hide. The weeds come back, and seeds often fail to germinate. After weeks of weeding and cleaning up. We can walk away content, believing we have done everything that we could to cultivate beauty. In a week, as we sit to enjoy our beds, one weed will catch our eye and then another. If we focus, we will see an entire army of weeds standing erect ready for battle.

Even gardeners wish the battle would end and that we could stop the world to get off. However, it keeps twirling around the sun like a little child unaware of any problem. No control would be scary, but gardening gives us power over what we can change and peace in our efforts.

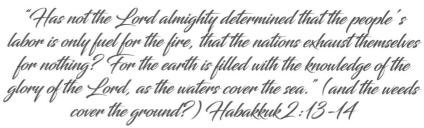

"Has not the Lord almighty determined that the people's labor is only fuel for the fire, that the nations exhaust themselves for nothing? For the earth is filled with the knowledge of the glory of the Lord, as the waters cover the sea." (and the weeds cover the ground?) Habakkuk 2:13-14

Gardening is not about any great individual day. Life like a garden comes into being because of small battles, constituting individual victories won, one day at a time. It is protecting our hearts from hatred, lust, laziness, busyness and the like. Gardening is guarding a place of refuge from reckless men or careless beasts: destructive like hungry bears or chewing beavers. They can be friendly like digging dogs or naïve of the harm they can do like chewing gentle chipmunk or undermining like burrowing insects.

Part of the way a garden is preserved is through canning, making pies, muffins, crisps and preserves. It is the harvest of our labors. Their conservation enhances life's flavors and sweetness. It makes it possible to remember them, taste them, and draw from them when we most need them. Gardening preserves us. It keeps a childlike wonder stirring within us. Though the same flowers bloom around us every year, gardening finds delight each time they do. The rewards of our investments change us exponentially.

"Your kindness will reward you..."
Proverbs 11:17a

The second part of preserving is restraint. When something is growing, it changes and gets bigger. Within a garden, pruning and trimming are quintessential methods of restraint. Many plants root from cuttings. Transplant and division in a garden preserve beauty and pass it on. A gardener then has blessings to divide and share. Division multiplies the blessings. This is contrary to how most of the world works, but exactly how God works. His is the source of increase, fruit and beauty. The blessing of God's hands all multiply in this way.

"You, God, are my God, earnestly I seek you: I thirst for you; my whole being longs for you, in a dry and parched land where there is no water. I have seen you in the sanctuary and beheld your power and glory. . . I will praise you as long as I live, and in your name I will lift up my hands. I will be fully satisfied as with the richest foods; with singing lips my mouth will praise you. On my bed I remember you; I think of you through the watches of the night. Because you are my help, I sing in the shadow of your wings. . ." Psalms 63:1-7

• Name something outside of the garden that increases with division.

• Describe a monster found in a beautiful place.

• Name something grown by gentleness and affection that force would stifle.

Watching and Observing...

God may not have meant for our lives to be so hard, but I know He always intended them to be prolific. When God made us gardeners He gave us reproductive abilities and gifts for creativity. He put a desire within us to accomplish great things. He gave us power and influence. Yet He limited our knowledge and understanding. He made us servants.

"Each one should use whatever gift he has received to serve others, faithfully administering God's grace in its various forms. If anyone speaks, he should do it as one speaking the very words of God. If anyone serves, he should do it with the strength God provides, so that in all things God may be praised . . ." 1 Peter 4:10-11

Sherlock Holmes said, "The world is full of obvious things which nobody by chance ever observes." Mastery requires time. Understanding is gained by studying an hour or a day at a time, and continuing in and out of seasons and over years and decades of time.

In life, as in a garden, there are ideal times for weeding, planting and pruning. The attentive gardener studies watches and takes no small thing for granted. Watching takes eye contact and focus. Silence observation exposes veiled detail learned in no other way. A scholar appreciates a subject of study more because a teacher challenges them to spend time looking and revealing layers of things. Other's observations will always broaden our own. In examination, we gain understanding and with understanding influence. Surveillance searches back and forth, over and over. It may look for signs of danger or good, but it also looks for timing.

Ugly corners of the world change, because someone took time to study them and develop beauty in them. The same thing can happen in relationships and people when time is taken to understand. Too often shallow observations judge our choices. When we look for beauty in a person or place, especially when it is not readily observable, we garden a relationship. We give the relationship and a person time to grow and a desire to reveal what is hidden. Caring for a person can reveal God given gifts in them we never could imagine. But when cared for, a person gathers courage and a desire to share their gifts.

C.S. Lewis uses fecundity to define life within a garden. Fecundity is the power to be fertile. It is not a measurement of what a man has produced but an understanding of what he is able to produce. When men speak of their humanity, they speak of their weaknesses. But God made us gardeners when he made us humans. Before the first sin was committed, He granted us great power, an ability to be prolific, and did not take that power away.

"For the Son of Man is going to come in His Father's glory with His angels, and then He will reward each person according to what he has done."
Matthew 16:27

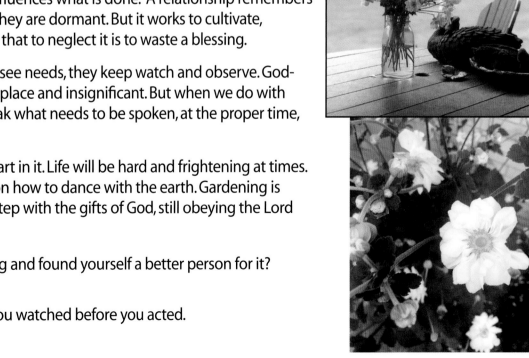

The measure of a harvest and fruit of one's life is not always an accurate gauge for understanding. Some things are harder to grow than others. Understanding influences what is done. A relationship remembers beauty when it is hidden and cultivates a person when they are dormant. But it works to cultivate, because it has observed good in them and understands that to neglect it is to waste a blessing.

Good servants are always looking to meet needs. To foresee needs, they keep watch and observe. God-given gifts are like seeds. They can seem small, common place and insignificant. But when we do with them what needs to be done, in the right place, and speak what needs to be spoken, at the proper time, they can make a momentous difference.

Gardening changes the world, because it changes our part in it. Life will be hard and frightening at times. We can watch as weeds overrun it or learn by observation how to dance with the earth. Gardening is ultimately about a relationship. To garden is to keep in step with the gifts of God, still obeying the Lord who set them in motion.

• Have you ever worked to watch and observe something and found yourself a better person for it? Blessed because you did?

• Share a time when you held back and were glad that you watched before you acted.

What Can Be...

Things are not always what they seem. Even in the best of conditions, living things can struggle for no apparent reason. They grow in wrong directions and need to be trained in new ones. They get sick and need extra care. Like the hydrangeas in front of our old farmhouse, if cared for, pruned, and given a second chance, God can do amazing things with unexpected places and seemingly insignificant lives.

Gardening breaks up the ground's curse. With God's hands guiding ours, gardeners sift through it and break it apart. God's power over the cursed soil is what matters.

Living things and persons are thriving, green, or blooming all around us. Others go unnoticed, are dormant, or going to seed. They can slowly yellow or wilt quickly before our eyes. Gardening is the pursuit of what can be.

There is never a reason not to try again. Some things die after the first frost, but others, like tulip bulbs need a cold snap to grow. In a brush fire most things die, yet some plants thrive because of them.

"What strength do I have that I should still hope? What prospects, that I should be patient?" Job 6:11

The strength and hope we have is in God's gift of a garden and its secrets.

Gardening begins with the end in mind. A child only gets to play in an enchanted garden when someone has planted a place ahead of time for them. Sore backs, scraped-up knees, and wind-burnt skin generates future places of beauty and moments of joy.

A garden is planted one year for the next. If cared for by one person, it is a gift for the next heart and life it will encircle. Gardening is the battle with what our garden is and what we hope it will be. It is growing fruit from faith that a seed is able to produce when it meets right conditions. It is power to do with ordinary things, small though great things of value. "Ordinary things consistently done produce extraordinary results," – Keith Cunningham. Because small deposits of time, over time redeems your time. (Unknown author)

Gardening is planting seeds and believing in future beauty even when no light or water is available. It is hoping that when it rains again, the ground finally thaws, or the sun breaks through the darkness new beginnings are possible.

"Hope doesn't disappoint, because God has poured out his love in our hearts. . ." Romans 5:5

Our Creator's unparalleled ability to look at what is, as what it can be, makes Him the Gardener John 15:1. Our belief in His ability to change the world will make us better gardeners.

• Have you ever come to value a relationship that you never thought possible upon first impression?

• Can you describe a time when after being pruned or wilting someone you knew came back with more vigor and strength? Has a brush fire or frozen terrain ever brought about a time of new growth in your life?

• Have you ever sought after deserted or neglected lives within your community?

• What are you cultivating with the end in mind?

Gardening invites God's kingdom back into our place in this world. As we work for what we know can be, we won't mind so much what is. We may only catch a glimpse of it for a moment, day or season in time, but His kingdom has great power and permanence. Fragrances permeate and rings ripple farther than we can ever hope to see. The Bible tells us, Moses and many others never saw what was promised. But it did not keep them from believing it. Hebrews 11:8-17

A Battle Won...

A friend sent me a card with a saying by Arthur Smith. It said,

"If you want to be happy for a short time, get drunk; happy for a long time, fall in love; happy forever take up gardening."

A profound thought.

It is good to be in love, but love remains love only so long as it is actively nurtured by both parties. And happy only happens when our relationships and our sense of accomplishment support each other. Gardening gives us the ability to return joy to those we love and watch their joy multiply. It produces true wealth. This joy of our own making surpasses the intoxication of wine and elation of first love.

"You may say to yourself, 'My power and the strength of my hands have produced this wealth for me.' But remember the Lord your God, for it is He who gives you the ability to produce wealth.
Deuteronomy 8:17-18

There is no tool available to man more powerful than love and no instrument more necessary than love for our fruition. Love waters before the heat comes and weeds despite thorns. Love gives us the desire to work despite biting bugs and gnats in our eyes all to make a place of peace for another. It is an unyielding belief in what God can do with sacrifice. It is a battle we choose very day to take up or neglect to accept.

Power and strength will never be enough. If our reason to rejoice is tied to anything less than God's abilities, stability will wax and wane as seasons pass. When life feels like the work is too much, we are trying to do what only God can do. When we refuse to replant after loss, we place the worth of what is lost over God's value in our lives. Like pretend garden fairies, we live in a false luminescence of our own making. The joy of gardening is bound to be snuffed out. If our hope is in God, who gives it all and can restore it, we can rest in His sovereignty.

When He is the ultimate treasure, our heart rests. As in Matthew 6:21

"Where your treasure is, there your heart will be also."

God gave Adam a place to love, before He gave him someone to love to ensure that he did not seek treasure from a person that only comes from cultivating God's will. Gardening is treasuring up, as in memory, the earth, a garden, and people. It is ultimately a battle to cultivate gifts within them and around them.

Gardening teaches us that we have very little control over the simplest things. Sin changes everything. Paying attention to things we do have control over will take our focus off hopeless battles.

Studying a garden has taught me that we should not be trying to change or manipulate those from whom we should be learning. Plants can be divided and yet not diminish. They multiply in division. A cutting taken from a plant mimics all the characteristics of its parent. They are rooted to one place and yet can produce seeds which can be carried around the globe. They are simple and green one day and then cloaked in colorful blossoms the next. God's hand cannot be mistaken in their glory.

Gardening is a battle that ultimately will be lost in this fallen world. As soon as they are abandoned, gardens begin to disappear. Our lives here have endings. The kingdom we cultivated, the God we honored, and His pleasure are the only treasure worth cultivating, and sacrificing for. Haggai encourages

". . . Give careful thought to your ways. You have planted much, but have harvested little. You eat, but never have enough. You drink, but never have your fill. You put on clothes, but are not warm. You earn wages, only to put them in a purse with holes in it. This is what the Lord God almighty says; 'Give careful thought to your ways. Go up into the mountains and bring down timber and build the house, so that I may take pleasure in it and be honored,' says the Lord. You expected much, but see, it turned out to be little. What you brought home, I blew away. Why?' declares the Lord Almighty. "Because of my house, which remains in ruin, while each of you is busy with his own house. Therefore, because of you the heavens have withheld their dew and the earth its crops. I called for a drought on the fields and the mountains, on the grain, the new wine, the oil and whatever the ground produces, on men and cattle, and on the labor of your hands." Haggai 1:5b-11

The battle we will lose, Jesus has already won. Without our Creator, gardening is futile. When He is our delight, all of our desires will be fulfilled in His answers.

"Delight yourself in the Lord and He will give you the desires of your heart." Psalms 37:4

• How has the joy you provided another changed you? How do you cultivate such a gift?

• Have you been bitten up by bugs, wind-burnt, frostbitten or sunburnt to cultivate a gift in another?

• How are gifts we get from relationships different than rewards we earn from cultivation?

• Have you avoided a relationship, because it was too much work? When is this a good decision?

GARDEN
Always

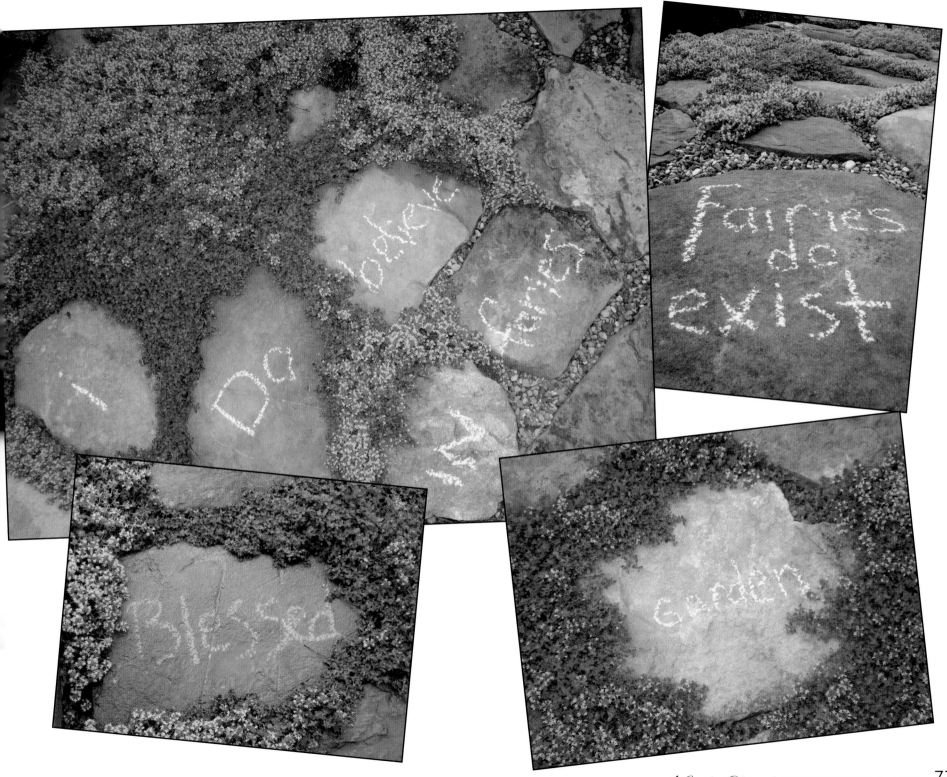

Concerncing Garden Keepers...

My daughter told people I was a garden keeper when she was growing up. I have often wondered what that meant to her. We are all garden keepers caring for different parts of the same garden. Some find fulfillment cutting the grass. Others living in skyscrapers struggle to care for houseplants. But we all cultivate something, be it friendships, children, and callings. In our jobs, relationships and private lives, we plant seeds, pull weeds and produce seeds or fruit of some nature. Some of us focus on flowers, some fruit. Each garden emanates a sweet-smelling fragrance or a stench.

Ezekiel tells us God made us to be beautiful jewels in the garden. Chapter 16 warns us that our choices can change what develops and grows.

"I made you like a plant of the field. You grew up and developed and became the most beautiful of jewels . . ." Ezekiel 16:7 a

My kids are both teenagers now. It is hard for me to imagine that God made them to be gardeners. In their right mind, no one observing our family would describe them as such. They love fast moving, competitive sports. Chores are beneath them. Their iPods and cellphones mesmerize them. They pick fights constantly. Relationships foster their worries. Like most teens, they honestly believe the world revolves around them.

In truth, my husband and I now turn and clear the soil. We encourage growth, by offering tastes of countless pieces of the earth's sweet fruit. We endorse their love for the Lord with as many bouquets of God-given riches as we can gather. We plant things they love and fruiting things we hope will someday sustain them. We weed constantly to keep sin from strangling their dreams. We name evil and talk about its characteristics as seedlings and as grown manifestation of that root. We plant new seeds, consenting to the fact that most will lie dormant for years. We water their hearts after wild, scorching days at school. Sometimes we utilize a light sprinkler; other days a soaking hose and on occasion a bucket directly overhead. We fertilize carefully. Some things we do not wish to encourage. We guard their foundations and ask them to be careful with elements they have yet to value. We have prepared and encouraged them in a seedbed. But obviously it is not enough. In their early teens, they are bored to death most of the time. Our house is a plot of ground they are rooted to and despise. Like the rest of us, they will begin to garden when they choose to change their grounding.

The Master Gardener has a purpose for their lives. Though their potential may not be apparent right now, I trust that the Lord will fulfill His purpose for them. He never gave up on me, and He will not abandon them.

"The Lord will fulfill his purpose for me; Your love, O Lord, endures forever Do not abandon the works of your hands." Psalms 138:8

As an acorn is encoded to become an oak tree, we are all small treasures in God's hands. With God as our gardener we stand, tall, branch out, and become exactly what He dreamed our lives to be.

"You brought a vine out of Egypt; you drove out the nations and planted it. You cleared the ground for it, and it took root and filled the land. The mountains were covered with its shade, the mighty cedars with its branches. It sent out its boughs to the Sea, its shoots are as far as the River. . . Psalms 80:8-9

- What tastes of heaven and bouquets of God-given riches do you share with your family?

- What things are you planting now to sustain them in the future?

- Do you have trouble accepting that some seeds lie dormant and only germinate in the future?

- What keeps you from seeing young people as acorns and believing that they can be oaks?

Needing Light and Water...

"The Lord took the man and put him in the Garden of Eden to work it and take care of it."
Genesis 2:15

God's clear intention for Adam before the fall, was that he would work the soil; that he would be a friend to the earth, stir up beauty and create and maintain a sanctuary. That Eden needed care really surprised me when I thought about it. I had imagined that place self-sustaining, but Eden needed a gardener. Because, "Nothing comes from nothing. Nothing ever could." – Sound of Music

I don't know what work in Eden may have looked like. Something tells me each day Adam woke, excited to begin. I know each night he slept sweetly. Pouring your heart into anything you love and watching it flourish ignites joy and purpose even in this day. "The place God calls you is the place where your deep gladness and the world's deep hunger meet." –Fredric Buechner. I am not sure Eden had weeds. What I am sure of is that it needed a gardener. Eden needed Adam. It should make sense that if any good would grow in our lives today, we are gardeners who must sacrifice for it. It needs gardeners willing to grab hold, dig in and draw up dormant beauty waiting in seeds for the warmth of the sun.

"Make it your ambition to lead a quiet life, to mind your own business and to work with your hands, just as you were told, so that your daily life may win the respect of outsiders and so that you will not be dependent on anybody." 1 Thessalonians 4:11-12

As a man or woman forms a garden, the garden forms a gardener. Without a garden's help a gardener ceases to exist. Repairing age-old foundations and ruins of a garden, mends the broken walls and clears the neglected paths of the one sacrificing for it.

"Your people will rebuild the ancient ruins and will raise the age-old foundations; you will be called Repairer of Broken Walls, Restorer of Streets with Dwellings." Isaiah 58:12

The world will always need paths that lead to sanctuaries, safe havens and refreshing pools. Those paths form as gardens are tended. A gardener will take into account that it matters what we tend and where our passions lead. Others will come behind.

"For the waywardness of the simple will kill them, and the complacency of fools will destroy them." Proverbs 1:32

Light illuminates. Light dances. It shimmers, flickers and brightens. It brings warmth and highlights elegance. But it also celebrates simplicity. It sheds, kindles, reveals, and guides. A cold world welcomes its gentle caress.

"Praying christian becomes the dream of God. He becomes the prism through which God dispenses his spectral colours to the hideous world." C.S. Lewis

Winter reveals how light's diminished presence affects the world. In summer, we bathe in it, yet sunburn discloses the delicate balance of exposure. Dusk and dawn herald it departure and arrival. It is central to survival.

Individuals bring light or take it away. We have all been in the company of those who make the world a different place. Some make us miserable in our happiness and others infuse our hearts with joy in the darkest of times. Shadows melt as the promise of their presence moves across our life's garden. Their reassuring dedication binds up the broken and brings light to the darkness. Jesus' light was just such a gift.

"He has sent me to bind up the broken hearted, to proclaim freedom for the captives and release from darkness for the prisoners." Isaiah 61:1

A gardener's life is like light to a garden. Immersed in the Creator's goodness, they are able to be content with the little they have. But they can dream of more because their hopeful endurance penetrates the dark canopy of the world and awakens its wonders. A child's giggle, an angry man's smile, or lifted eyes of a grieving heart are remuneration. Gardeners remember those moments when November's chill takes away summer's warmth.

". . . (A gardener, like) Love never gives up, never loses faith, is always hopeful and endures every circumstance."
1 Corinthians 13:7-8 a

Faithfulness will reveal countless scratches, an occasional bruise, and sweat soaked affection. A gardener's love overlooks hurt and engraves joy on the tablet of their hearts.

"Let love and faithfulness never leave you; bind them around your neck, write them on the tablet of your heart." Proverbs 3:3

I have a friend who laughs easily and ignites joy all around her. I call her my sunshine fairy. She is always in the sun whether she is working or resting. I like to relax and labor in the shade. I tend to dwell on darker things when I am not writing and focusing on God's gifts. She helps me to laugh in spite of myself. She radiates light. I absorb as much of her company as I can. Not every plant grows best in full sun. Some like the shade and develop best in the understory, and we know those rays of light are precious.

I consider some of my friends cool-breeze fairies; they offer excitement and refreshment. Some are sparkle fairies. They glisten in the shadows and light. One most precious to me is a water fairy. She savors parts of the world others miss and teaches me to soak up life, jump in and bathe in its waterfalls. Our differences help us appreciate our need for one another. I cherish every one of their sweet souls and need each one at different times. They add blessing to my life I do not deserve, yet accept wholeheartedly.

"Some people look for a beautiful place, others make a place beautiful."
Hazart inayat Khan

[li]fe is fragile and changeable. The most basic things it needs are [w]ater and light. I needs each as much as the other.

[W]ater is reflective, but also active. It can glisten. It thunders, drips, [tr]ickles, and streams. It can saturate, permeate and seep through [cr]evices not so easily penetrated with conscious effort. Water is [tr]ansparent, tasteless, and odorless. Yet in times of severe drought, [it] is possible to see, feel and hear the earth longing for it. It is [u]ndervalued until we are desperate for it. Too much can be just as [d]eadly as too little. It can extinguish the light and yet without it, [th]ings burn up in the sun. That balance seeps into life in ways we [h]ardly recognize. The earth is a desert we are destined to travel [lo]oking for a well or a source of light (like a sunset), as in The Little [Pr]ince by Antoine de Saint-Exupéry. Best found in community and [c]ompanionship of souls that match our own.

[A]s a garden flourishes when it is watered before the sun is hot, [re]lationships depend upon gentle affection. An affectionate [g]ardener is a cool drink of water to the wilting.

[T]he magnitude of our dependence on a simple thing like water [sh]ould awe us. Not every garden can have a fountain, stream or [po]nd. Maintaining the algae and bacteria in a pond can be extra [w]ork anyway. But every garden needs a source of water. Every life [d]epends upon a fresh spring.

Jesus said, "If you only knew the gift God has for you . . . you would ask me and I would give you living water." John 4:10

God uses gardeners to water a thirsty world. When He withholds the rain, He gives us the power to offer ever-flowing hope, joy and peace that finds it source in Him. John 7 promises that, when we believe in Him, living water will flow from us.

"Whoever believes in me, as the scripture has said, streams of living water will flow from him." John 7:38

Our lives can be as gentle as the morning dew, refreshing as a waterfall and restoring as a cool drink of water to a wilting world. As our words and influence can be scorching and burning, our lives can also be as destructive as a flood, stagnant, and dry as a rock.

"He opened the rock, and water gushed out; like a river it flowed through the desert. For He remembered His holy promise..." Psalms 105:41-42

• Do you presently work for something that helps you to sleep sweetly and that you wake excited to cultivate?

• Is there some part of your life or hobby that you enjoy that partly defines you?

• Are you leading others to follow after your passions?

• If you were a garden fairy what would you want your name to be?

• Have you ever come out of a dark time and realized how valuable someone's commitment was to your survival?

• Has the light from a friend's life ever brought back joy and laughter for you?

• Do you need different friends at different times in your life? Why did God make us this way?

• Do you find that knowing someone cares for you and pours out some form of affection is uplifting? Bonds you to them or separates you?

Just a Gardener...

"God put us first into a garden and when we lost eden we were fated to search to recreate it and find it again. Only some of us know how to do this."
A Little Chaos the movie.

I had not always, but I have now come to love calling myself a gardener, with all the grim and glory the name conjures. Like Eve before me, in my self-importance, I failed to see the significance of gardening. I thought there were more important things to do and be. I wanted control over more than the soil.

"You made him ruler over the works of your hands; you put everything under his feet." Psalms 8:6

I recognize now, if my life is to be a blessing, I must be a servant to everything under my feet. Dirt clings to me; my nails break and hands crack; my hair is windblown and my skin is sunburnt. I can't stand up straight by the end of the day, but restoring a broken and barren world is more important than a life of ease and self-promotion.

When work leaves a gardener haggard, it can be easy to notice the garden and miss the gardener. A gardener was overlooked in the Bible. In John 20, Mary made this blunder.

"At this she turned around and saw Jesus standing there, but she did not realize that it was Jesus. "Woman," He said, "Woman, why are you crying? Who is it you are looking for?" Thinking that He was the gardener, she said, "Sir, if you have carried Him away tell me . . ." Jesus said, "Mary." . . .
John 20:14-16

When Mary heard the Savior's voice as she walked away from an empty tomb, I don't think it was insignificant that she thought Jesus was just the gardener. I think she assumed He was the gardener, because the gardener was always tending the place. She presumed the gardener would always be there.

"It's only you," is so easily the thought we associate with those who have been a part of our lives through mundane and tough times. Recognizing, "It's always been you," about those same persons, indelibly etches their names in our memories.

Those who teach us our significance and give answers for our emptiness are God-given gardeners. They are the people who tend us when we are unable and help us grow. When we are children, they clear safe paths and cultivate safe havens. When we are teens, they weed a world that would consume us. When we are angry and hardened, they plow our fertile soil; they water and fertilize what is deep beneath the surface; they trim and prune what we put on display for the world. In their battle for us, gardeners will battle against us. They may make us angry. They can frustrate us as much as they love us. They confront what we allow to flourish when it needs to be trimmed back or even torn out. In pruning, they allow light to shine on important parts of our lives that we have yet to value. Gardening like…

"Love bears all things, suffers long, and believes all things." 1 Corinthians 13:4

These gardeners may be looked down on. In rebellious teenage years, it is to be expected as teenagers despise their parent's gardening. Those who garden us kneel and write forgiveness in the dirt. They get their hands dirty, haul away our mistakes, like weeds and battle our stubbornness with a shovel. Their focus is on encouraging other's fruit and beauty not their own. They challenge us to be all that we can be. They garden us in our infancy, in our adolescence and our grown life. In our deepest grief, driest, most mundane years and simplest joys, they are always there. Because we assume they will always be there, sometimes we fail to be awed by the truth that they are.

Perhaps, because she was looking for someone else, Mary saw The Gardener in her time of grief, though she failed to see her Savior. Most of us recognize the gardeners in our lives during turmoil when we find ourselves alone, in trouble and life gets scary; their voice calling our name is a source of comfort.

When we are struggling, we can fail to attribute the value we should to many precious figures. I try to remember this as my teens lash out and hide from me. I focus on loving them despite their thorns and rejoice over any fruit or blossom. I hope that by never giving up, they will learn to see me differently: my cultivation instead of my correction.

• When did you first recognize a gardener who challenged you despite your rebellion? How important was that person's influence?

• Looking back were you glad that they never gave up? Does this give you hope as you are now the gardener of others?

• Do you remember needing their voice for reassurance?

A Garden Observed: CULTIVATING A LIFE

A Gardener Defined...

If asked to describe a gardener, most people think of an old sun hat, weathered and windblown, gloved hands and nippers, scraped up knees, and dirty boots. But a gardener is so much more. Human beings, like countries, are "enigmas wrapped in mysteries enveloped in riddles." – Winston Churchill. Separated by decades and by thousands of years we are much the same. Fifteen-year-old boys and 50-year-old women need to be gardened. Ninety-year-old men and 4-year-old girls were meant to garden. Yet few people see themselves as gardeners today.

Consider why a gardener makes a garden: why a gardener tends to the grounds over a cemetery, around a church, business or home; why any gardener does more than the bare minimum to survive in the world? Why would anyone want to add more work to our exhausting lives? Gardeners are treasure seekers, artists, and visionaries.

A gardener's heart encourages growth. A garden keeper defines a line and cuts a perimeter allowing ground covers to grow so far and no further. He fills each vacant nook and empty cranny. He plants perennials for the future and with annuals brings color to the present.

A garden is a collection of treasures and by planting it a gardener is assured of being able to share it. Like Mary, Jesus' mother, a gardener,

"... ponders all of these things and treasures them in her heart (or garden)." Luke 2:19

As a child cherishes favorite items left by a deceased parent; or a parent cherishes items of a lost child; so a gardener gathers The Creator's gifts. A gardener runs his fingerprints through the water and across tree trunks. He or she traces the tangible impressions of the Creator's fingertips.

"Now faith is being sure of what we hope for and certain of what we do not see . . . By faith we understand that the universe was formed at God's command, so that what is seen was not made out of what was visible."
Hebrews 11:1, 3

My friend, Mary Wiley Lewis, is an artist. She once told me that when she paints a landscape, she first figures out what is important. She forgets the clutter and paints what is left.

A gardener uses the soil to plant, as an artist uses paint to paint, a place worth gazing upon, spending time and sharing with others; a place worth remembering. He drags away all that is rotten and cuts away what is overgrown. In this mixed up world, sometimes it is hard to sort through the clutter.

"Different kinds of gifts, same spirit, different kinds of service, same Lord, different kinds of working, same God in all."
1 Corinthians 12:4-6

God sees what we will be, though we are only what we are. What defines a gardener is the ability to see what something or someone can be despite dirt or jungle. It is a gentle strength that wields beauty from understanding.

"A bruised reed He will not break..." Isaiah 42:3

Has this book helped you to define what a gardener is and does? How will you change?

Do you showcase and highlight what gifts God has given you? What do you now collect? What harvest from them do you share?

Kneeling...

Consider the story of Ahab and Jezebel 1 Kings 21:1-28. Naboth was murdered, because Ahab was not a gardener. Ahab may have owned the garden of Naboth after Jezebel murdered him, but Ahab was never its gardener.

Naboth's garden was no longer a place of refuge. The memory its name invoked was never the same. Ahab thought that as a king, a garden should be readymade. He was not willing to kneel in the soil or wait for seeds to grow. If he were a gardener, he would have created refuge, not taken one.

Sam Gamgie in Fellowship of the Ring said about the elves, "…no one could be sure if the land made them who they were, or they made the land what it was." So each man forms the land and the land forms him. A garden's keeper is shaped by its formation. A gardener will be different at different times of the day and certainly every season of the year. In season, he blooms, fruits and flourishes despite winter or drought. In times of transition, she may enter dormancy or wilt. Each spring he will bend down and encourage again.

"…You stoop down to make me great." Psalms 18:35b

We are all searching for some great purpose, but our purpose is simple. We are to plant and cultivate blessings. A gardener germinates when she loves something. He grows when he is willing to kneel down for it. Roots are put down when her life is stretched and he reaches out for water and light. His life is worth living when its purpose is worth kneeling, worth dying a little each day for. Love ignites passion and purpose in the heart of the one who wields them.

"If I cannot do great things, I can do small things in a great way." MLK

A gardener may not always be the person we think we need, but she will be the person there for us, when we are in need. They bring both beauty and food for hungry souls. Their work is their gift.

"He who works his land will have abundant food, but he who chases fantasies lacks judgment." Proverbs 12:11

I've seen so many movies lately about exceptional gifts granted to ordinary and exceptional individuals. I think there is a reason we only see it in movies. Too often parents are passionate about labeling children gifted and fail to train them to be a blessing. God doesn't give individuals extraordinary talents very often. Even when he does, the gifts are not for categorizing our life's purpose. They are for sharing.

"My heart is not proud, O Lord, my eyes are not haughty; I do not concern myself with great matters or things too wonderful for me. But I have stilled and quieted my soul . . ."
Psalms 131:1-2a

n the United States, God has showered us with gifts capable f creating extraordinary conditions and places. But in our lectronic age, we have come to want the garden readymade. /e see ourselves as kings. As Ahab, too many of us pout and el sorry for ourselves, and demand that for which others ave worked. We want miraculous lives and tend to look own on those with simple ones. We can't understand why e are unhappy. We pursue pleasure and power and believe ardening is beneath us. Instead of producing treasures and ifts, we can share, demand beauty and fruit from others, ping the land in our frustration. Far too few people have hosen to be gardeners or to change their own grounding.

Do not withhold good from those who deserve it, when it is within your power to act." Proverbs 3:27

ach life makes its own difference. Each has the power to cultivate good or evil, neglect or urture what he has been given. Each touches the lives of those around him. Gardener's ecognize the difference one life can make.

"Cursed is the one who trusts in man, who depends on flesh for his strength and whose heart turns away from the Lord. He will be like a bush in the wastelands; he will not see prosperity when it comes. He will dwell in the parched places of the desert, in a salt land and where not one lives. But blessed is the man who trusts in the Lord, whose confidence is in him. He will be like a tree planted by the water that sends out its roots by the stream. It does not fear when heat comes; its leaves are always green. It has no worries in a year of drought and never fails to bear fruit."
Jeremiah 17:7-8

• Have you seen others stoop down to make others great?

• Have you challenged your children to be blessings?

Plowmen...

We bought rescue horses. The one my daughter fell in love with had fear issues. The horse felt every emotion of everyone around her all the time. Emma's only goal in life became saving that horse. One afternoon my daughter and her friend took their two horses on a trail ride through the woods. On horseback, they attempted to cross a bridge. One horse's leg went through the planks and its hoof was injured. The girls called. They were afraid to walk back alone and wanted us to come to them. We had trouble finding them. So my daughter lead her horse toward the end of a field to meet us. Because Emma was afraid, they fed off each other's anxiety. By the time I reached her, the horse was in a total panic and Emma was shaking with fear. The horse wanted to run to the farm. She would not turn to go back to the other horse. Separating the farm and us was a 200-yard plowed field. The trenches were femur breakers not ankle breakers. It was cut through with ditches 2-3 feet deep. I had struggled through it to get to Emma. It was frozen and snow covered. If we let the horse go we knew she would plow through it, kicking and running. I knew the only way to save both of them was to let the horse go, but Emma had a death grip on her. I took the horse, and (I didn't need to see it), I could feel Emma's terror as I let her go. She kicked a few times, but quickly began slowly tiptoeing to the barn. She was not hurt. I followed the horse; Emma led the other mom back to her friend.

There are some relationships we need to let go of and times in them that we will have no control.

"... they have greatly oppressed me from my youth, but they have not gained victory over me. Plowmen have plowed my back and made their furrows long." Psalms 129:2-3

The more we love and depend on someone, the more they can hurt us. Even good intentions can be carried out wrongly. Relationships turn over life's soil. The best and worst of them leave cuts and scars. Others dig trenches and a few channels. Some cultivate.

Farmers plow to open up the earth. It allows the winter rains to soften it. As the snow piles up, it freezes and thaws breaking down and melting into a spring landscape ready to ignite with new life.

Relationships have a priceless ability to break through stagnant crusts of routine and ritual. Healthy ones empower us. C. S. Lewis said in The Four Loves, "We are helpless. As soon as we are fully conscious, we discover loneliness. We need others physically, emotionally, intellectually; we need them if we are to know anything, even ourselves." Those connections between people need continual gardening. For this a person plows to influence, disciple and exhort.

It is in our broken places, we are most often used by God. Caine

When a garden is first planted, the ground needs to be cleared. Fields need to be plowed. This is necessary seasonally. If it is done constantly, it destroys the ground and breaks up soil structure needed for root growth.

"Listen and hear my voice: pay attention and hear what I say. When a farmer plows for planting does he plow continually? Does He keep on breaking up and harrowing the soil? When He has leveled the surface, does he not sow caraway and scatter cumin? Does he not plant wheat in its place, barley in its plot, and spelt in its field? His God instructs him and teaches him the right way." Isaiah 28:23-2

Plowmen plow. But it's not always to hurt us and not everything that causes pain will harm us. In life, as in a garden, we have to choose what to do with the wounds, whether to constantly stir the soil or work through wounds. Walking over them compresses them, creating a path or channel. Those paths are safe if they lead to places of restoration. Avoiding deep wounds, like channels, will leave open pitfalls in our territory. They can be a water diversion. Streams form that can wash away topsoil.

"A despairing man should have the devotion of his friends, even though he forsakes the fear of the Almighty. But my brothers are as undependable as intermittent streams, as the streams that overflow when darkened by thawing ice swollen with melting snow, but that cease to flow in the dry season and in the heat vanish from their channels." Job 6:14-17

There is a chance to plant something new when a plow has torn apart an area of our lives. When we easily let go of earthly things, especially hurts and offenses, a plowed land can be replanted. Complaining or stirring up a place of hurt never allows restoration.

None of us are called to remain in a connection that constantly leaves deep furrows across our backs. But if given a winter's rest, spring to germinate and summer to flourish, harvest and beauty are possible from community. It is very easy to see how much we need the Lord when our lives have been plowed and we stare at brown vacant landscapes. A garden keeper knows a relationship with him is the only assurance of a rich soil with safe paths and deep roots. But not only that. A garden keeper depends on him for guidance in relationships so that they feed off of His peace not the anxiety around them.

• Does the energy of people around you affect your own feelings. Has anything helped you battle that (other than substance abuse).

• Has a relationship ever plowed through a stagnant crust in your life? Or hindered your growth?

• Though you were hurt share a time when a relationship helped you to root more deeply, branch out and bloom more easily.

Now Good To Be Alone...

When I was younger I thought that relationships would be easier as I got older. But I have not found that to be true. Relationships are complicated. Girls and boys are as different as men and women. When he was little, my son would have friends over, a fist fight would break out and 10 minutes later everyone was laughing about it and playing video games again. That never happens with little girls. A punch is never thrown. A word is never spoken, but a hurt can divide one girl from 14. A small offense can divide one group into many. The wound isn't forgotten easily. Some last a lifetime. In the teenage years and adult life, relationships are hard. The reasons are not always clear. Many gardeners seek solitude to remedy or avoid misunderstandings.

For a long time I gardened alone, not because I wanted to, but because that is where I found work. The longer I was alone, the darker the world became. I missed people, communication, and conversation. LIGHT. Burdens are lightened by those who work beside me. The nice thing about gardening with others is knowing that when a rock is too heavy, or wheelbarrow is too full, someone will be there to help me. My timing is better with help. I do not grow tired as quickly. Their stories lighten my heart; their shared troubles make them as human as me. My prayers are more powerful and purposeful as I see and feel their shared sorrows and memories.

"... and if you spend yourself in behalf of the hungry and satisfy the needs of the oppressed, then your light will rise in the darkness, and your night will become like the noonday." Isaiah 58:10

In the beginning God said, "It was not good for man to be alone." Our lives grow and build so much more in community than ever they could in isolation. Bridges and skyscrapers are never built by individuals. Beautiful places can be made by communities and can join diverse groups of people.

God made relationships, community, and friendship to make labor worthwhile. Years ago in the back of my Bible, I wrote this quote. I wish I knew the author. "We never lock eyes with someone who does not warrant an all out search. We never lock eyes with someone God does not long for. A price has been paid, an all out search is on for them. Without them something valuable wound up missing." Without someone searching for us beneath the jungle, after the hurricane or despite the drought, we would be lost.

Sometimes we share a place with strangers. Some strangers we will know our whole lives. We will never see what is beneath the surface. We cannot all be best friends. C. S. Lewis said in The Four Loves, "Friendship is not necessary for survival, but it makes survival worthwhile." Survival in a blank brown landscape is a given until God calls us home. Anything different than a barren plot of land has to be intentional.

If a gardener was brave enough to plant a favorite thing, would it change the place for everyone? If one person chose to water what was wilting, would it matter? If another weeded what was hurtful and called it by its name, would it keep it from scattering seeds? If someone fertilized what was struggling, someone gathered bouquets of flowers and deadheaded what needed rest, would everyone benefit? If everyone were nourished by the fruit that was grown in a cooperative effort, would not each benefit exponentially from the division of labor? What would the world look like if our time together was enjoyable and not just survivable? Peter teaches us to do this.

"Each one should use whatever gift he has received to serve others, faithfully administering God's grace in its various forms." 2 Peter 4:10

It is important to have a healthy balance in our work and relationships; between our place and the persons with whom we share it. With the Lord's help we find a healthy balance.

Jesus replied, "No one who puts his hand to the plow and looks back is fit for service in the kingdom of God.
Luke 9:62

Each person is called to plow his own soil,
plant his own seeds and care for his own gifts.

"Finish your outdoor work and get your fields ready; after that, build your house." Proverbs 24:27

But wisdom tells us we do not have to do it alone. When not being worked, a garden is a place to be shared. A gardener sets a stage for meeting. Flowers in vases remind us that our time together is special. Fruit can be used to make wine for sipping, grains for breads to soak up oils and herbs; berries with creams and countless other treats to savor in rest after hard labor. He prepares a place to build relationships, a place that refreshes, heals and bolsters our hearts. This only happens if we meet others with awareness and draw them with cords of loving kindness.

"The Lord appeared to us in the past, saying: "I have loved you with an everlasting love; I have drawn you with loving kindness. I will build you up again. And you will be rebuilt... Again you will plant vineyards... the farmers will plant them and enjoy their fruit." Jeremiah 31:3

Friendships are cultivated working side by side with others. Sharing the harvest is an espace and reward. Friendship is the place we flourish and flower.

Gardeners come with fruit when others are hungry. They bring flowers to the grieving, clear paths when someone is lost and define borders when lines need to be drawn.

Friendship unlocks the potential of both parties despite hardships. Gardener-friends bring transplants and share fruit. Their life's lessons can add to ours and keep us from harm. An invaluable friend shares the beauty and joys of our gardens. One who picks up a shovel, puts on gloves and gets on knees with us is more than can be hoped for.

"The pleasantness of one's friend springs from his earnest counsel." Proverbs 27:9b

Other gardeners know how hard it is to grow what we plant. They know how much the thorns hurt. They know the same dirt that covers our feet and hands, and what can breed from the filth if we do not wash it away. Jesus showed us what to do with that knowledge.

"Jesus knew that the Father had put all things under His power... so he got up from the meal... poured water into a basin and began to wash His disciples' feet, drying them with the towel that He had wrapped around Him." John 13:3-5

Some gardeners, as some friends, will be called to work all day and night to make a difference. Others will have little bits of time here and there to give. Take breaks to visit, wash each other's feet and savor those gifts as often as you can. Whatever time you have together, friendships are a gift worth investing in.

"Now that I, your Lord and Teacher, have washed your feet, you also should wash one another's feet. I have set you an example that you should do as I have done for you."
John 13:14-15

My prayer in writing this is that maybe someone will be inspired to kneel beside another, make eye contact and begin to cultivate.

A gardener is a common person, exceptional because of uncommon love and purposeful action. A gardener is different from the rest of the world and the world is a different place because of him. Their life is a garden with intimate treasures that overflow in the lives of others.

"You are my own private garden, my treasure, my bride, a secluded spring, a hidden fountain."
Song of Solomon 4:12

- Do you like working with other people? Why?
- Can you think of something that cannot be built without a community?
- What needs to be built by an individual?
- Have you seen the culture of a place changed because someone used a God-given gift administering God's grace to a community?
- Do you have a friendship or friendships that make survival worthwhile?

GARDEN *Always*

Concerning Soil...

My kids and our dogs constantly run in and out of my garden beds. And, I spend too much time in them myself. Foot traffic compresses the soil. It's a problem. To amend this and the normal settling soil, I break the soil open with hand tools and chop up hard clumps. It invigorates the soil.

A rare person ever notices the reason for lush elements of a garden. Most other people simply see me scratching at the ground like a chicken and politely say nothing. It will never be my favorite chore. (And I have yet to eat bugs as a practice). But it is a priority, because I observe its benefits. Sometimes the best way to tend to what is growing above the surface is to tend to what is beneath the surface.

The circumference of a plant's canopy collects most of the rain water. This circumference is called a drip line. Sometimes you can see it. When soil between each plant and around this area is loosened air and rain showers penetrate easily. Undulating surfaces hold fertilizer in place when applied and gives it time to absorb. When this happens roots reach out and plants flourish.

Soil... soul.

"Don't be tired of doing good. Don't get discouraged and give up, for we will reap a harvest of blessings at the appropriate time."
Galatians 6:9

What takes root and fills life's territory, what causes life to be common or uncommon is influenced by the soil from which it draws strength and anchors roots. Some are mired by this truth. Some claim dominion over it. God governs both noble and common standings. Hence, very different things grow from the soil of each man's life.

". . . Shall what is formed say to Him who formed it, 'Why did you make me like this?'" Does not the potter have the right to make out of the same lump of clay some pottery for noble purposes and some for common uses?" Romans 9:20b-21

The things a man neglects, compresses, or buries within the soil of his heart affect the depths of his being and fruit of his life. Every so many years a farmer may allow a field to rest, but unplowed ground in life often exposes what we have no courage or are too lazy to change. Showers wash blessings away, blessing that God meant to saturate our lives.

"Land that drinks in the rain often falling on it and that produces a crop useful to those for whom it is farmed receives a blessing of God."
Hebrews 6:7

As a garden's potential is governed by the soil, a man's life is influenced by his foundation. Climate, weather and geographic location affect a garden. Those things are, for the most part, out of man's control. The soil is not. The first step, to amending an old or creating a new garden (a new purpose for life), is to pay attention to what God has written in the soil beneath our feet.

"They were using this question as a trap, in order to have a basis for accusing Him. But Jesus bent down and started to write on the ground with His finger." John 8:6

What unseen or seemingly silly things do you do to invigorate your life?

Is there unplowed soil in your life?

How could you benefit by examining your foundation?

Dirt...Cultivated Soil...

"He has thrown me into the mud. I am nothing more than dust and ashes."
Job 30:19

From the time we are children, we are all trying to figure out what magic makes the mud different from our own hearts. For most of us the journey begins when our parents start trusting us around mud puddles. It's a big deal when people begin to trust our judgement around dirt, especially the sloppy kind. But one day it happens, and they let us down in the backyard. Though they watch to see we are safe, they stop worrying that we will eat the mud again. As we grow, they are not so worried that we will run and jump into the sky reflected in puddles on the ground. Our interest concerning mud puddles changes as their trust in us grows.

"For now we see only a reflection as in a mirror; then we shall see face to face. Now I know in part; then I shall know fully, even as I am fully known." 1 Corinthians 13:12

There are a few men in the Bible remembered for their connection to the soil. Uzziah is one of them. He was remembered because ". . . He had people working his fields and vineyard in the hills and in the fertile lands, for he loved the soil." 2 Chronicles 26:10b

The Lord gave him success so long as he sought Him. But when Uzziah became powerful and prideful, he was afflicted with Leprosy until the day he died. 2 Chronicles 26:16-21 a

Uzziah's passion was a gift. The opportunity to make a living from it was a double blessing. When we cultivate them for the Lord they can prosper and sustain us. When we focus on the gift and forget the Giver, everything that mattered is lost. The rich soil becomes dirt with all its despised qualities.

"When pride comes, then comes disgrace. . ." Proverbs 11:2

The difference between soil and dirt is foundational. Horticulturalists never call the ground beneath their feet dirt. Dirt is swept up and thrown out. Kids drag mud in after making footprints in puddles. After the ensuing mud-throwing battle, it gets itchy. A snake's curse is that it crawls on its belly and eats dirt Genesis 3:14. It has frustrating and unwanted qualities. Dirt is a foul or filthy substance. It has contemptible and mean qualities. Man covers his head in it in times of remorse as in Lamentations 2:10. Dust, with dirt-like qualities, is seen in the Bible in Lamentations.

"The Lord in His anger has cast a dark shadow over beautiful Jerusalem. The fairest of Israel's cities lies in the dust, thrown down from the heights of heaven. In His day of anger, the Lord has shown no mercy even to His Temple."
Lamentations 2:1

The dirt flies up and hits us in our face as we weed. It is where we lay in our grief (Job 2:8). Our bodies are abandoned in it when the earth beneath us falls away. The dust or dirt is associated with man's abasement, lamentation and frailty. The disciples shook the dirt from their feet as a sign of rejection (Acts 13:51). David (2 Samuel 16:13) and Paul (Acts 22:23) had dirt thrown at them.

We plant in the soil not in the dirt or the dust. It is important to know what differentiates basic things that can be easily misidentified and are abundantly present in our everyday existence.

"Who appointed Him over the earth? Who put Him in charge of the whole world? If it were His intention and He withdrew His spirit and breath, all mankind would perish together and man would return to the dust." Job 34:13-15

The dirt of the world covers vast landscapes with a dust that offers no foundation. It can root us into jungles where we become so lost we fail to recognize that we are.

Pride hardens our hearts as the sun and foot traffic can cement our soils. Pride will often not consent to change. When pride hardens our hearts, breaking that pride can be problematic. We need more than hand tools to revive it.

"I will break down your stubborn pride, and make the sky above you like iron and the ground beneath you like bronze. Your strength will be spent in vain, because your soil will not yield its crops, nor will the trees of the land yield their fruit." Leviticus 26:19-20

There is dirt and there is cultivated soil. Life comes from soil."

The Lord formed the man from the dust of the ground and breathed into his nostrils the breath of life, and the man became a living being." Genesis 2:7

The Bible also talks about the earth with soil-like qualities in many verses. You will see a difference in the meaning of the dust, ground, earth or field associated with soil. Soil has God's blessing forming its purpose.

"From the dew of heaven and the richness of the earth, may God always give you abundant harvests of grain and bountiful new wine." Genesis 27:28

"Faithfulness springs forth from the earth, and righteousness looks down from heaven." Psalms 85:11

" But finally the earth is at rest and quiet. Now it can sing again!" Isaiah 14:7

To a gardener, soil is land we long to own. It is living, deep, rich earth when it is cleared of all but light and water. It is nothing like dirt. Soil is beautiful, black and loamy. It flakes and floats in our fingers. A healthy soil, as on a forest floor here in Pennsylvania, is usually dark, has large and small clumps and is made of clay and small particles like sand. Built by plant materials, minerals and nutrients from the building blocks of life, it encourages life.

Good soil is a rich mixture of structure, porosity, and pH. Large and small particles give it both sand-like and clay-like qualities. Water is absorbed for future use and slowly released with nutrients as roots ask for them. It allows water to filter through, refreshing it easily. Its porosity does not consent to stagnant conditions which suffocate growth. Air pockets in it allow roots to push through without being hindered. It sustains microorganisms beyond number which must be kept in healthy numbers. Life is kindled in its embrace.

It is the soil, not the sun or seed, which determines the long term health of a garden. Its physical properties reveal why one garden flourishes and others struggles. The same root stocks and the exact amount of light needed can touch every corner. A seed and the sun begin to uncover its value after one spring shower and time reveals the depths of its merit. The truth about its inner parts cannot be hidden. The same is true of men and women. Wisdom comes from the soil of our hearts.

" Surely You desire truth in the inner parts; You teach me wisdom in the inmost place." Psalms 51:6

Soil acts as an anchor. Living things germinate and establish themselves within it. It can hold a massive oak tree firm and secure. Yet it still encourages new beginnings when the smallest seed is covered in it. The soil of the land is where roots push out and blessings come into being.

"Those the Lord blesses will inherit the land, but those He curses will be cut off." Psalms 37:22

Across the globe, the land can be vastly different. In Pennsylvania, fertile fields and thick forests stretch out for uninterrupted miles. The land in the Middle East changes drastically from one area to another. A lush and fertile vineyard can be a few miles away from a barren desert covered in rocky caves.

The soil each person anchors his or her heart can be as varied as the number of individuals in a room. Men and women with a thick rich soil have the potential to offer abundant harvests. However, some survive in a thin layer over the surface; no root penetrates deeply. Others have been trampled underfoot for so long their hearts are compacted and no seed has the opportunity to germinate.

The parable of the four soils is found in Matthew 13:3-9. Different soils are contrasted in it. In Crazy Love, Frances Chan warns, "We must be careful not to assume we are good soil." In the explanation of the parable in Matthew 13:18-23 understanding is indicative of good soil.

Cultivated soil is plowed annually. Rocks are pulled from it. Even after a hundred years of farming, farmers here in Pennsylvania still spend the spring pulling rocks from their fields. A couple times a year they amend it with manure and spray it for weeds. For a deep rooted relationship with our God, we will need to plow what is dormant, pull up rocks that surface, weed what is evil and fertilize what needs to be nurtured.

Then the Lord stretched out His hand and touched my mouth, and the Lord said to me, 'Behold, I have put My words in your mouth. . . To pluck up and to break down, To destroy and to overthrow, To build and to plant."
Jeremiah 1:9-10

The soil from which a person draws nourishment can also be drastically different from one place in life to another. My dad was a coal miner. I know my childhood was a healthy one, because he poured out the best part of his youth deep beneath the ground for his family.

"All that is gold does not glitter,
Not all those who wander are lost;
The old that is strong does not wither,
Deep roots are not reached by the frost".
JRR Tolkien

The roots of my childhood were anchored by his labor in the darkness. The soil's secrets envelop all that is hidden: seed, stone, boulder, hope, harvest and treasure. Since the beginning of time, the soil needed a gardener. What is buried there anchors relationships and fosters choices. It matters if it is rich or poor, thick or thin. But what can be done with it depends more on the master of the garden than on the soil's apparent characteristics. In the hands of the right gardener, diminutive things can rise up and great things can be established.

"He raises the poor from the dust and lifts the needy from the ash heap . . .
For the foundations of the earth are the LORD's upon them He has set the world." 1 Samuel 2:8a, c

• Can you separate the dirt from the soil of your life?

• Once separated, can you identify how to fix your foundation? Are you willing to make a change?

• What helps you to hold on and let go, filter life, send down roots?

• What anchors your roots and encourages beauty in you?

• Do you confront the hard patches, pull stones regularly? Depend on regular fertilizer to survive?

Clay and Sand...

love being a mom. I am crazy about my kids. Looking back I remember being physically exhausted while my kids were small. But surviving teenagers is a different kind of tired. It emotionally draining. Forbidden fruit is always more appealing than what I have to offer. arents understand that its consumption can destroy all that we have worked to cultivate. ur hearts can collapse like sand and harden like clay as we wrestle with God through arenting. Only prayer and trust in God's grace keeps us from turning into toxic pillars of salt ke Lot's wife. We have to make a decision not to keep looking back at what we should or ould have done differently. (Genesis 19:26)

fe batters a person in the same way rain, sun and time beat the earth. Life pummels us, ress bakes us and our busy schedules compress us. People walk all over us.

"For hardship does not spring from the soil, nor does trouble sprout from the ground. Yet man is born unto trouble as surely as the sparks fly upward." Job 5:6-7

I suspect the soil held many of Eden's original characteristics up until Noah's flood. It probably slowly changed as weed seeds blew; it was lived on or abandoned. I am sure flood currents swirled it around mountains. Every downpour, fire, wind storm, lightning strike and bulldozer since continues to influence it.

Watch a heavy downpour assault loose earth and the curse is almost observable. Mud splashes, water puddles and streams form and run down slopes. When it stops raining, tiny pieces bond together and melt into mud. If it is never cultivated, layers pile one on the other.

God called the dry ground good, but the original blueprint He provided in Eden is no longer a part of the land's promise.

"Cursed is the ground because of you; through painful toil you will eat from it all the days of your life. It will produce thorns and thistles for you and you will eat the plants of the field. By the sweat of your brow you will eat your food until you return to the ground . . ." Genesis 3:17b-19a

I think God made working the ground painful after the fall to make sure we were paying attention. Our every action influences our foundations. Sin will make our hearts like clay if it is not softened with His love, and like sand if we are not fearful of His power.

Soil-like characteristics surface in all of us. Sandy soils typically lack nutrients, and clay soils hold them like a clenched fist. Some soils cannot be penetrated. Others lack stability.

At times, clay is reluctant to let go of the water molecules. But when it is waterlogged, it creates a vacuum that suctions anything dug up right back into place. It even sucks tractor tires down into its grasp making it unworkable and tractors unmovable. When clay dries out, however, it does not soak water up. A shovel bounces off of it and gardener's joints are injured in a continual battery. A pick axe works best in these conditions. If it is hard for a shovel to penetrate clay, tiny roots of things we want to encourage struggle exponentially more. Clay is slower than other soils to warm in spring.

Sandy soils struggle against the curse just as much. Life struggles to establish itself in it and nothing can be depended on from it. It heats up quickly in the sun and can scorch the feet of those who walk over it. Sand has trouble holding water and nutrients. Both are obviously vital for plant life to thrive.

Men and women, like the clay, can be impenetrable holding onto and controlling everything around them. Reluctant to let go of the past, we all hold captive what could be useful. We can be dependably unusable and hard to change. It frightens us. Insecurity, pride, embarrassment and regret waterlog and vacuum us back into old habits and ways that are hard to break. These things keep many worthwhile things from growing under its care. Those who work with us are injured incessantly digging through our obstinacy. My husband and I have been in a few arguments where a pick axe would have worked wonders for both of us. We stifle our own growth and that of those to whom we cling. Clay-like hearts have the same effect as the chains in.

Psalms 107:10. "Some sat in prisons and deepest gloom, prisoners suffering in iron chains."

old seasons in life are hardest on those of us with an iron clasp on the past. We can be coldest to those closest to us and our frozen demeanor can diminish reluctant growth despite gentle showers of affection.

ike sandy soil, our lives can leach away every good thing life offers. We can fail to notice or remember what is important. People can e both reckless and careless about the cost of their choices. If constantly living for the moment, we are weak like the sand. We fail to repare for times of drought and stress or recognize how fragile life is. Psalms 103 reminds us.

. . . his days are like grass, he flourishes like a flower of the field; the wind blows over it and it is gone, and its place remembers it no more." Psalms 103:15-16

Jesus warned us never to build on sand.

"But anyone who hears my teaching and doesn't obey it is foolish, like a person who builds a house on the sand." Matthew 7:26

he curse of the ground also makes it either acidic or alkaline. When moss is growing over the surface of a grassy yard, it is acidic nd called sour here in Pennsylvania. Adding lime to the soil sweetens it. The grass is greener, thicker, and the weeds less numerous hen the soil is sweetened. When we fail to see what is important and shut others out, our gardens are left desolate and our soil ours.

"Even small children are known by their actions. . ." Proverbs 20:11.

Zechariah illustrates how people can affect the soil they influence.

"They made their hearts as hard as flint and would not listen to the law or to the words the LORD Almighty had sent by His Spirit . . . When I called, they did not listen; so when they called, I would not listen, says the Lord Almighty. I scattered them like a whirlwind among all the nations, where they were strangers. The land was left desolate behind them that no one could come or go. This is how they made the pleasant land desolate." Zechariah 7:12-14

When the fruits of the spirit ripen in our lives, our soil is sweet. But even when the land has a healthy structure, it grows whatever takes root. It harbors incalculable weed seeds. It is held down by extensive root systems unintentionally encouraged. The autumn forfeits growth. Freezing and thawing cause the land to heave and collapse. The burden of winters stills the land and melts in the spring. Summer dries it. It is never satisfied by water and is stagnated by too much.

The earth is a desert, tundra, a swamp we are destined to travel looking for soil, looking for souls with which to put down roots and change the our world. If the soil is to be changed and useful someone must cultivate. If we are to be changed, we will need communities, friendships and souls to cultivate us.

• Do you battle a clay or sand like foundation most of the time?

• Have you had the opportunity to garden in different soils? How are they different?

• Have you seen a gardener sweeten a soil you once thought sour?

• Are you apart of a community that cultivates your soil?

Plowing...

he best thing that could have happened to my marriage was my parent's move. All of my siblings had moved on by the time they sold the rm and moved out of it. I needed mom when my kids were small, and ad her all to myself. She was my closest friend. However, I spent long urs with her and leaned on her and dad instead of my husband for mfort.

hen they moved, all that was familiar, my home, favorite walks, and mily were suddenly far away. I had to cling to my husband. I had to d new paths, make an effort to know my neighbors and look to my usband to fill the void. I cried on occasion. But small efforts and planted eds began to grow. I branched out as I embraced my neighbors and rapped my heart around my husband's.

hen our lives are plowed, starting over is hard work, but asserted effort is not a bad thing. In fact, the hard things in life have the eatest potential to form us. The fiercest battles most of us fight take place deep in the soil of our hearts. When God confronts what onstitutes a man's heart He often uproots, tears down, and destroys what keeps us from Him.

".... Can I not do with you as this potter does?" declares the Lord, "Like clay in the hand of the potter, so are you in my hands, ... If at any time I announce that a nation is to be uprooted, torn down and destroyed, and if that nation repents of its evil, then I will relent and not inflict on it the disaster I had planned. And if at another time I announce that a nation is to be built up or planted, and if it does evil in my sight and does not obey me, then I will reconsider the good I had intended to do for it."
Jeremiah 18:5-10

Change is hard. Even if we don't like things they way they are, we are accustomed to them. We know what to expect from them. Rejoicing over alterations takes courage. When things change there is a time of upheaval. In a garden, there is a time of plowing.

"You have shaken the land and torn it open; mend its fractures, for it is quaking."
Psalms 60:2

The purpose of plowing is not to wound or hurt the land. The plow's blades uncover what is hidden and lay it bare to revitalize it. Plowing opens a weed-bound, hardened and stagnant layer of fertile terrain. New life is able to rise through its broken surface.

What a life is capable of growing depends intimately upon the hindrance polluting it and the soil encouraging it. Our soil is the part of ourselves we keep from God. It is the part of ourselves we will not allow him to fill. Until we consent to being plowed, it is a part of our lives that we guard deeply or reveal in intimacy. When God cuts through the soil, He breaks the barriers of our hardened hearts. God pounds (cuts, plows, irrigates, aerates, cultivates) gently. He digs up roots of bitterness; He plows Satan's lies under our feet. The change begins deep beneath what others observe.

He knows it is very hard to establish something new battling weeds and root systems undermining His purpose. So he lays everything bare before us.

"Nothing in all creation is hidden from God's sight. Everything is uncovered and lay bare before the eyes of Him to whom we must give account."
Hebrews 4:13.

C.S. Lewis explains best why God does this, "We are not in the business of cutting down jungles but of irrigating deserts." God plows under the jungle so that He can water just what is most precious in us.

Meeting the Lord with hand tools or a spade is one way to approach Him. At times God shows up with a tractor and His own plow. At other times, for His own reasons, we meet Him in a drought, brushfire, or rising waters.

"It is a dreadful thing to fall into the hands of the living God." Hebrews 10:31

Any of these options cutting across what has rooted within us is frightening. When God, the Gardener plows, it will be the wrestling match of our lives. But we will walk differently. Jacob was changed in this way (Genesis 32:22-31). We must be willing to be injured and not let go of Him.

He always clears a place for blessing. The Lord transplanted the Israelites from Egypt for this reason. He planted them in a land He cleared so that they could fill it with His kingdom. Newly cleared ground may look like a desert, but when God is the one filling it an oasis, promised land, and Eden are its promise.

"The Lord will guide you continually, giving you water when you are dry and restoring your strength. You will be like a well watered garden, like an ever flowing spring." Isaiah 58:11

• Can you share a time of plowing in your life? Did you find your heart lain bare before the Lord?

• How did you regain you foundation, form new paths or make new friends?

• Were you permanently injured? Did something new and more precious grow within you? Have you walked differently since?

Honesty...

I think the truth of our lives is equal to the soil of a garden. We are all born with an amazing sense of honesty. Children bellow it. Parents are quick to teach children not to hurt others feelings and not to be too honest. But, a child is amiss when he or she begins to see hidden truths as impotent and to withhold and avoid truth themselves. It is important to teach appropriate place to dig into truth and appropriate times to examine the soil. Even for adults it can be hard to know when to break up hard ground, plant something new and trust the Lord to send showers.

"... I will drive (you), (You) must plow, and (you) must break up the hard ground. Sow for yourselves righteousness, reap the fruit of unfailing love, and break up your unplowed ground; for it is time to seek the Lord, until he comes and showers righteousness on you. Hosea10:11b-12

I really struggled with the truth openly spoken to God in Psalms 137. It was a chapter of the Bible that I had to ask the Lord to give me peace about.

"... happy is the one who repays you for what you have done to us - he who seizes your infants and dashes them against the rocks." V.8b-9. In Psalms 137,

We see the sorrow of a person dealing with the harshest truths. In this chapter, we hear the writer's heart poured out in the presence of the Lord. He shares a raw honesty when he asks the Lord to repay hurt that he was helpless to stop as a captive. Perhaps he helplessly watched as horrors were committed against his mother, wife, sister, or any young woman... We hear his hurt and anger and watch him pour it out in the Lord's presence. We can see his body shaking as he weeps beside the river in his honesty. The man exposed his heart's bitter soil before the only One who could restore it. God desires the same exposure from us.

"He does not treat us as our sins deserve. Or repay us according to our iniquities. For as high as the heavens are above the earth so grea is His love for those who fear Him; as far as the east is from the west, so far He has removed our transgressions from us. As a father ha compassion on his children, so the Lord has compassion on those who fear him; for he knows how we were formed, He remembers that we a dust. As for man, his days are like grass, he flourishes like a flower of the field; the wind blows over it and it is gone, and its place remembe it no more. But from everlasting to everlasting the Lord's love is with those who fear Him and his righteousness with their children's children." Psalms 103:10-17

God is not always plowing, but He always uses confession to refresh His people. A place is cleared for healing in transparency and vulnerability. Confession rejuvenates us. It pulls the roots of things that sap our strength and darken our hearts. Confession can be hard. Jeremiah explains why.

"The heart is deceitful above all things and beyond cure. Who can understand it? I, the Lord, search the heart and examine the mind, to reward a man according to his conduct, according to what his deeds deserve." Jeremiah 17:9-10

Confession reveals to us what is hidden. If we understand our Heavenly Father's true nature, we recognize that He already knows everything. All the walls we can build and fig leaves we hide ourselves behind will not cover the naked truth.

"He reveals the deep things of darkness and brings deep shadows into the light." Job 12:22

Confession pours out and breaks open our most appalling secrets. Some things are best left unsaid before man. But anything less than childlike honesty before God is to withhold, restrain, and avoid truth. It leaves things in the soil that need to be confronted. What is severed can then be mended.

"Come, let us return to the Lord. He has torn us to pieces but He will heal us; He has injured us but He will bind up our wounds." Hosea 6:1

When our longings lie open before the Lord, He removes what should be and renews what is lying dormant and fertile.

"All of my longings lie open before you, O Lord; my sighing is not hidden from you. My heart pounds, my strength fails me; even the light has gone from my eyes." Psalms 38:9-10

Preparing the soil in the garden, reminds me to be honest with myself and to be honest with God. Honesty is a catalyst for healing.

The Bible says Noah was a man of the soil in Genesis 9:20. After the flood, I have to ask: "Did he have anything else to build on?" Today the same question is pertinent. Do we have anything else on which to build? Anything worthwhile that produces beauty and blessings anchors itself in the truth. A garden is not a stagnant thing that can be overlooked, put in a vault or buried under the ground. (Matthew 25:14-30 Parable of the Loaned Money) Our lives always grow something. What truth it is anchored in affects what it is capable of growing. How it is cared for and encouraged then influences its harvest.

Some people lack a moral compass and other's rule with insensitive judgment. Some people lack stability and are constantly needy. Others keep their hearts hidden and refuse help even when they need it. But what composes a person's heart affects what he or she is capable of producing.

"You have plowed wickedness, you have reaped injustice,
You have eaten the fruit of lies. . ." Hosea 10:13a

If our hurts are to be healed and our longings are to be answered, truth must expose them. Honest exposure allows righteous things to be established.

Hebrew reveals that the truth can hurt.

"For the word of God is living and active. Sharper than a double-edged sword, it penetrates even to dividing soul and spirit, joints and marrow; it judges the thoughts and attitudes of the heart." Hebrews 4:12

The Psalms tells us to keep our eyes on the Lord through the hurt.

"They will say, "As one plows and breaks up the earth, so our bones have been scattered at the mouth of the grave. But my eyes are fixed on you, O Sovereign Lord; in you I take refuge - do not give me over to death." Psalms 141:7-8

Unnamed doubt, anger and fear do not mean that it does not exist. Giving them names gives us power over them. I am not trying to draw people out to share a raw honesty that wounds or is used to stir up strife. But honesty will identify the soil, the very ground on which we stand before our God. The Truth provides a Light that leads us to the Way.

"Jesus answered, 'I am the Way, the Truth and the Life . . .'" John 14:6a

". . . He (Jesus) said, 'I am the Light of the World. Whoever follows me will never walk in darkness, but will have the Light of Life.'" John 8:12

It may seem dishonorable to approach our Heavenly Father with the crude realities of our choices and our thoughts. I know. There are things I struggle to be honest with Him about. But, He is our only hope. He will not desert us or continue to recount sins of our confession. He alone can restore beauty from trite ashes and remove pollutants that seeped through stagnant soil. The best way to guard our hearts is to expose them before Him.

"Above all else guard your heart for it is a wellspring of life." Proverbs 4:23

He wants it all. God is not injured by the truth even if we were. He is not afraid of what we fight. He will shatter illusions and lies. Then gently, but definitively, turns over and cuts through what should never have rooted. At that time restoration can begin as we see in Zechariah.

"The seed will grow well, the vine will yield its fruit, the ground will produce its crops, and the heavens will drop their dew. I will give all these things as an inheritance to the remnants of this people." Zechariah 8:12

• Can you think of a time when a child said something that was correct, yet it should not have been said out loud?

• How was the situation handled? What did the child learn?

• Have you ever wrestled with bitter sorrow and honestly and then felt its weight lifted before the Lord?

• Have you been transparent and vulnerable with a person who then became your friend? Lost a friend because of it?

Timing...

Roses don't bloom in the winter, at least not in Pennsylvania. I am sure it has rarely happened in the past. Yet I remember a single rose blooming in mid-January at the old farm house. In the warmth of summer, a stem had grown between a crack in a storm and interior-window. So that while blizzard like winds howled outside and feet of snow covered the ground, a solitary bloom exposed its red pedals between the two windows. I still don't know how it was possible, but my family remembers it too so it wasn't a dream. It was like the enchanted rose in Beauty and the Beast.

Timing in a garden, as in life, is often understood by seasons. In spring, showers "drench furrows and level ridges. (God) soften(s) (the ground) with showers and bless(s)" what is planted within it. (Psalms 65:10) In mid-summer, most gardens are in their glory. However, beginning a garden in the middle of the summer is risky. It is not impossible, but it is harder. The ground is as hot and dry as the air. Temperatures then cool in fall. As plants enter dormancy, the soil hardens inch by inch. Of course, frozen ground will not yield for growth or beauty. Shovels break and plows scrape across the earth's crust.

"There is a time for everything, and a season for every activity under the heavens:" Ecclesiastes 3:1

In life countless things depend upon timing. Yet time can run out. There can be too much time and not enough, bad timing and a good time. Gardens reiterate this truth. Farmers have to take advantage of whatever window of time spring opens for them to work the soil. If it is missed, a crop can be lost. In life as in a garden, it is best to take advantage of every opportunity that presents itself. We should not hesitate to plant whenever we are able.

C.S. Lewis says, nature does not teach, but nature will help a person understand the meaning of things that cannot be understood without it. Spring, summer, fall and winter come to all of us. Fall and winter can be harder to understand in life's calendar. But God can use times of dormancy and depression to prepare us for spring and vibrant growth.

Transplanting is best done when deciduous trees have yet leafed out in spring or after they drop their leaves in fall. Digging does minimal harm at this time, because plants are either coming out of dormancy as the soil warms or entering a state of dormancy as the soil cools. As the weather changes and transplants begin to grow it sends out only what leaves it can sustain. The plant will have time to anchor deeply before branching out.

Yet despite weather and seasons, God can thaw the soil beneath a frozen exterior and unexpected beauty can flourish and unfold in life. In God's hands flowers bloom despite freezing temperatures. His love is a greenhouse that can reach a heart even in the icy sting of sorrow.

God can do surprising things with the harshest seasons of our lives. Sometimes what He asks of us can seem as impossible as getting a rose to open in the frozen ground. After the harshest of seasons, Maple trees flow with a sweet sugar. They remind us that God uses every season for good things. Maple syrup is only harvested for a short window of time in spring. If He can bring sweet sap to us from a tree, He is able to harvest any life's purpose in His timing.

We will make mistakes. At the farm, some things we tried to save gave up and some transplants did not survive. Failure did not keep us from trying time after time to make things right or at least better. When I think back about the disrepair that it was left in when we first moved in, I wonder if

the owner, previous to us, was unhappy. Lonely and broken lives often look like unkempt gardens. I wonder if no offer of help ever came or if it was not accepted when it did. The house took on the physical qualities of neglect either way.

Knowing when to cultivate a relationship, plow, plant, or weed and when to wait can be pivotal to its longevity and enjoyment. Sometimes people need to be ready to receive us and sometimes God can cause a friendship to grow despite harsh seasons. At times, those conditions are what make a friendship possible.

Gardens physically bloom, branch out and cascade with beauty because someone helps them to do so. Relationships require the same affection.

God's timing is perfect. His love is a greenhouse that can be depended upon in any season.

"But do not forget this one thing, dear friends: With the Lord a day is as a thousand years, and a thousand years is as a day. The Lord is not slow in keeping his promise, as some understand slowness. He is patient with you, not wanting anyone to perish . . ." 2 Peter 3:8-9

God would not that any blessing or beauty in our lives would be missed. He has buckets to be gathered filled and with the syrup of pure, lovely, admirable, excellent and praiseworthy things if we trust Him.

"Finally, Brothers, whatever is true, whatever is noble, whatever is right, whatever is pure, whatever is lovely, whatever is admirable - if anything is excellent or praiseworthy - think about such things."
Philippians 4:8

• Has anything miraculous ever happened in your life at the wrong time?

• Has nature ever helped you understand what nothing else could?

• Have you ever seen a harsh season of life bolster a person instead of breaking them?

Amendments...

My husband and I are very different. He is light hearted and laughs easily. He is quick to take on new adventures without worry. I plan for what can go wrong. My goals often take away my ability to savor the moment. We have been married so long now that our soil is well blended. We know how to argue and laugh with each other, and at each other. It helps the world to roll off our shoulders. When we want to it is easy to block out everything and seek out the best parts of life together. But that was not always the case.

His ways of doing things often made me fearful. My way of doing things often left him confused and agitated. Our first few years of marriage were tough. He was starting a career and his life was fast paced compared to mine. When he came home he needed to relax, but I wanted to escape. It took a while, but in time his strengths mended my weakness. My strong points fortified his weak points. The mixing of our soils was not easy, but the asserted efforts we made to stay together paid off.

Amending is the removing of faults. Yet when the soil is amended, things that are lacking are added to it. The addition improves it.

Sand or wood ashes added to clay-like soil helps with drainage and diminish the clay's swamp-like qualities. Clay added to sandy soil helps with structure which helps it hold together and retain water. It also adds nutrients. Organic matter in the form of compost can be added with either clay or sand. And sometimes, soils need all three to be added. It enriches the soil immeasurably. As a gardener identifies what the soil needs when he or she amends it. God knows each heart and He knows what each needs. Yes, He weeds. He clears the stones. Then, He amends the soil. He adds to our lives. He brings people to our lives that do just that. As our soils blend, it can seem like a hard transition. He does not reduce them. He enhances whatever potential is lying beneath the surface.

" But let your adorning be the hidden person of the heart with the imperishable quality of a gentle and quiet spirit, which in God's sight is very precious."
1 Peter 3:4 ESV

A balanced life holds loosely to what is precious and releases it when it is time. It allows offenses to roll off its shoulders and absorbs the reign of God's grace. All the qualities of a balanced foundation keep fear from crippling us and our hearts from being hardened.

We will meet clay, sand and compost in other people. Relationships enrich our lives. That does not mean they are always easy. As our lives are plowed and turned over by the world our soils will mix. If the truth is exposed and we are honest, our lives are amended because of what our friends teach us, our spouse offers us, and even what our foes bring out in us. The mixing affects our grounding. When given time to grow, battles with and for each other, shared hardships will intertwine our hearts. Laughter will then trellis what blooms because hardship improved it.

This turning over of the soil of our souls before the Lord is the beginning of a garden. After it has been torn up and mixed, after it has been planted, it will need protected. Mulch is a protective barrier and benefits the soil in many ways. It keeps the loosened earth from splashing up on stems, foliage and surfaces. Mulch insulates roots from the bitter temperatures and shelters it from the heat helping it to hold moisture.

In the garden, mulch lovingly covers the scars left from working through our gardens. It covers offenses as love covers our mistakes.

"Love (like mulch) covers a multitude of sins."
Proverbs 11:17

As arduous as change may be, when we first sip the wine from the grapes we harvest and eat fruit from the seeds we started, the labor pains will instantly be forgotten. We will not dispise the land we care for. Our lives will be fertile ground, sanctuaries in a lost world.

"The days are coming," declares the Lord, "When the reaper will be overtaken by the plowman and the planter by the one treading grapes. New wine will drip from the mountains and flow from all the hills. I will bring back my exiled people . . . they will rebuild the ruined cities and live in them. They will plant vineyards and drink their wine; they will plant gardens and eat their fruit. I will plant (my people) in their own land, never again to be uprooted from the land I have given them." Amos 9:13

• What was added to your soil by another that amended your heart?
• Have you ever had someone confuse and frustrate you only to make you a better person after understanding them?
• How could your soil be amended?
• Make a list of everything you would like to remember if a time of plowing ever changes things. Collect seeds, transplant…

GARDEN
Always

Concerning Seeds. . .

The time I spent in the gardens at my childhood farm taught lessons far removed from the activity. I was unaware of this truth at the time. But, I think memories influence us very much like seeds influence a garden. Seeds planted within us in childhood grow into the blossoms we love and weeds we hate in our adult life. But if you want to meet the flowers you're going to have to wrestle a few weed.

Memories are things we hold dear and try to forget. We share some and never speak of others. It is really important that we have them. Bad memories can be as valuable as good one. Historians keep records of them for a reason. They shape our past and influence our future. They make some storytellers, others gossips. They help us to relate to or to reject each other. Like caterpillars, some are worth getting to know.

"Well, I must endure the presence of two or three caterpillars if I wish to become acquainted with the butterflies. It seems they are very beautiful."
- Rose from the Little Prince by Antoine de Saint-Exupéry

Seeds can look like nothing more than a bit of dirt. They are cloaked in coverings that fail to reveal the purpose for which they are created. Most of us, given a handful of random seeds and asked to describe them, would not recognize what we hold. Microscopic plants hide in their coats. Appearing harmless, helpless, and insignificant, those hard coverings guard storehouses of potential. Seeds can seem inconsequential. However, since Eden, they are fundamental.

*". . . He made the things we can see and the things we can't see . . .
created through Him and for Him." Colossians 1:16*

Man was formed from the earth; and where there is soil, there is seed. However, all seeds are not the same. A seed's heart echoes its source though it is silent. The essence of something is hidden in the seeds that spill from its flowers. The seeds of beautiful flowers may not appear much different from the seeds of the most antagonistic of plants. But, they are kernels of different things. When they are misidentified or mishandled, the environment around them is effected. Some sprout and produce juicy, sweet fruit; others grow poisonous berries. Some seeds rise up to produce wind blocks and hedgerows; others are planted for roots systems capable of preventing steep slopes from eroding. Some grow into thorn patches and others into healing herbs. Seeds create havens of escape filled with smells like heaven and flowers for every season. Others produce plants with putrid aromas that attract flies and cause death.

In any given piece of land, as in any given person, there is an extraordinary amount of energy lying dormant beneath the surface. Hard labor is a curse man battles in large part because of the uncalculated number of seeds accidentally and intentionally planted. Some lie dormant, while others sprout quickly.

Learning to sort treasure from trouble is a valuable skill. Not even a microscope reveals all truths. Fruition alone reveals color and dimensions programmed in DNA.

"No good tree bears bad fruit, nor does a bad tree bear good fruit. Each tree is recognized by its own fruit. People do not pick figs from thorn bushes, or grapes from briers. The good man brings good things out of the good stored up in his heart, and the evil man brings evil things out of his heart. For out of the overflow of his heart his mouth speaks." Luke 6:43-45

God's original design for the world was His kind of good. He provided a perfect place. Since Eden, each man planted and dropped seeds that have changed God's design to man's own version of good. We have altered His blueprint. The ground's curse hasn't shown signs of diminishing.

"But you have planted wickedness, you have reaped evil, you have eaten the fruit of deception..." Hosea 10:13

The Bible clearly warns of two things:

*"Those who plow evil and sow trouble, reap evil and trouble."
Job 4:8*

*"He who sows seeds of wickedness reaps trouble. . ."
Proverbs 22:8*

All things grow to be what their Creator intended them to be. It is so important for us to discover the secret of what is growing and harness its energies. Good things in this world do not come into being by mistake. Good things, planted in a man's heart and his life, are no accident. Even the smallest acts and choices we make have great power over our lives. We don't need to be time travelers to change the future, the choices we make everyday change the future right now. It may even heal the past. Take notice of and time to identify the seeds scattered with every decision we make.

"Sow for yourselves righteousness, reap the fruit of unfailing love, and break up the your unplowed ground; for it is time to seek the Lord, until He comes and showers righteousness on you." Hosea 10:12

• Do you know any caterpillars who became butterflies – other than your own family members?

• How have particular memories shaped your life?

• Have you misidentified or mishandled seeds that you later identified? How would you now handle them differently?

Seedbed...

everyone of the overflowing love of our Father. Sitting areas were included so that we would remember to spend time talking and sharing stories with each other.

Memories keep us together and those we have lost are remembered in shared stories, not plantings. Though I doubt anyone who planted the trees around the building will forget digging the holes and placing sapling in the ground. I hope their children will share stories with their grandchildren of how their grandparents planted that tree at the church. Yet nothing in the landscape was spectacular in the first year. However, the following spring as the seedlings grew and perennials raised their stocks, beauty quickly drew every eye to canvases of color celebrating our love for our church.

A seedbed is a place of growth or source development. Gradual manifestation is encouraged in one. They can be intentionally developed. What is capable of expanding, enlarging, and branching out of such places is important to realize.

Support groups can be like seedbeds. They can appear to be blank, boring pallets, but for individuals with common hardships, small glimpses of beauty break through the soil. There are support groups for widows, those who lost children, cancer survivors and even dieting individuals. I was part of one while my children were small. Raising them was easier for me because others shared their experiences in child rearing. Shared grief, frustration, and hardships built bonds that are not easily uprooted. Kindnesses found in them offer water and light from those who know how much and when it needs to be given.

Seedbeds can manifest out of areas of devastation or desperation. Whether what grows is good or bad depends upon what seeds are encouraged. The silent stillness can be an awesome opportunity to find water and the warmth of healing and supportive relationships.

y church built a new building at a new location! We chose a design, broke ound, and poured a foundation. Walls rose plank by plank. The ceiling is eld up by carefully preserved beams cut from aged trees. Windows admit ys of light. Sidewalks guide visitors indoors. A green lawn encircles the rking lot, and brown canvases of soil were left for planting. The building anged. Our church and its memories grew. Seedbeds lay waiting to be anted around it.

ent soil beds presented an awesome opportunity for creativity. They ere also intimidating. So many things could be done. Congregational embers wanted favorite trees and flowers planted in memory of loved nes and special occasions. It was hard to honor all those ideas and high pes. So anchor trees and shrubs were chosen first. Then, we filled up e open spaces. Plants were chosen that complimented each other and eir repetition harmonized the property. Unity was the focus, and formal ds that invited children and backdrops of beauty were cultivated. disappearing fountain was placed outside our main door to remind

"For as the soil makes the sprout come up, and a garden causes seeds to grow, so the sovereign Lord will make righteousness and praise spring up . . ." Isaiah 61:11.

God puts us in unhindered places where the soil is soft and pliable at times. "He brought me into a spacious place because he delighted in me." 2 Samuel 22:17. He does this so that our cloaks can be shed and roots can dig deep until we are ready to stand on our own.

"In days to come you will take root, and your family will bud and blossom and fill all the world with fruit." Isaiah 27:6

In this place, the light from his care is gently introduced. The wind, harsh world and scorching stress are blocked by his love. His living water is the source of survival. He protects us from invasion from crueler things than birds, pests or pets. He surrounds us with others who fertilize us with gentle compost. Only the gentle, natural goodness of compost safely encourages development. Harsh salts of manufactured fertilizers burn tender roots.

"Peacemakers sows in righteousness."
James 3:18

Gardens begin with dirt that has no perceivable beauty. Seedbeds are not always obvious. They start void of life, but they do not stay empty for long. Just as a seed can grow unexpected things, so God can cause things to grow in the least likely places. Our lives will not return void when we watch over, pray over and nurture what God plants.

"For my thoughts are not your thought, neither are your ways my ways," says the Lord, "As the heavens are above the earth, so are my ways higher than your ways and my thoughts than your thoughts. As the rain and the snow come down from heaven, and do not return to it without watering the earth and making it bud and flourish, so that it yields seed for the sower and bread for the eater, so is my word that goes out from my mouth: It will not return empty, but will accomplish what I desire and achieve the purpose for which I sent it." Isaiah 55:9-13

- Can you look back on life and recognize seedbeds that you saw as awesome opportunities?
- What did you do with the open landscape?
- Do you have any memories tied to a tree or a garden you planted with someone special?
- Have you ever joined a support group hesitantly and then embraced it as a treasure?
- Have you known any peacemakers who protected you in times of growth?

Planting...

Our lives always plant something. What we plant in others manifest in their emotions. Our lives cultivate confidence or insecurity. People may forget a name, but they never forget how a person makes them feel.

I remembered meeting my dad's favorite aunt once. There were six of us, and because mom and dad were outnumbered, they rarely took us visiting. But dad had learned that she was sick, and he wanted us to meet her.

I was in junior high and I loved that visit. I decided when I grew up I was going to be like her. She sat down with us, asked about our lives and tried to make us laugh. She told stories and gave us homemade cookies. She died shortly after that. She had to have known she was dying and maybe was hurting and afraid. But she made us feel important and special despite her pain.

I also remember people who treated me very differently. My opinion of myself was wounded and twisted by the seeds they pushed into the soil of my young heart. It took me a while to uproot their hurt. I hope nothing about me is like them.

"Again He said, "What shall we say the kingdom of God is like, or what parable shall we use to describe it? It is like a mustard seed, which is the smallest seed you plant in the ground. Yet when planted, it grows and becomes the largest of all garden plants, with such big branches that the birds of the air can perch in its shade." Mark 4:30-32 Parable of the Mustard Seed

Characteristics of the mustard plant and seed are noted in this verse. The plant is grown for its seed. In most other places this plant is an invasive nuisance. This may seem offensive, as a description of the kingdom of heaven, but it testifies to the fact that it will never be eradicated and is not easily subdued. Seed from the plant multiplies easily. Its unrelenting habit of sneaking back into plowed soil makes it a weed. When planted intentionally, it grows tall and thick, out contending other plants and becoming a bush.

While many memories are never shared with others, it does not mean they are forgotten. The smallest seed planted has great influence over the garden as our smallest memories can have great authority over us. The ones we have the hardest time trying to articulate and share are often the ones we spend our lifetime trying to understand. I am sure that is why we are warned.

"Be careful what you plant." Genesis 9:20

Seeds are planted by habits: good and bad, consciously or unconsciously. They may be planted on purpose or out of ignorance. As parents, we plant confidence or insecurity. We nurture obedience or disobedience. Our lives bring forth both life and death. We plant both blessings and curses.

"This day I call heaven and earth as witnesses against you that I have set before you life and death, blessing and curses. Now choose life, so that you and your children may live and that you may love the Lord your God, listen to his voice and hold fast to Him. For the Lord your God is your life, and He will give you many years in the land He swore to give to your father's . . ." Deuteronomy 30:19-20a

Most seeds are planted in a hole and then left to grow on their own. The Giver of Life, alone, gives them the power to become what they are meant to be.

" . . . What you sow does not come to life unless it dies. When you sow, you do not plant the body that will be, but just a seed, perhaps of wheat or something else. But God gives it a body as He has determined and to each kind of seed He gives its own body. All flesh is not the same. . . The body that is sown is perishable; it is raised imperishable, it is sown in dishonor, it is raised in glory; it is sown in weakness, it is raised in power; it is sown in a natural body, it is raised in a spiritual body." 1 Corinthians 15:36-39, 42-44

In all of us, God plants seeds. They are small kernels that mark the beginning of great change.

"Do not dig up in doubt what you planted in faith." - Elisabeth Elliot

When your kids are teenagers you watch their every mood and move, as we pray over them and weed around them. We see seeds of greatness beginning to grow. This is an important time to put all our hopes in the Lord. Until they are established, they are on the verge of triumph or failure. The smallest seed sprouting in his heart, choices he makes or voices he listens to can change and affect him. We should be prayerful and careful about our part especially when things go wrong. We have to trust God to restore them and establish their foundation.

"I planted the seed, Apollos watered it, but God makes it grow. So neither he who plants nor he who waters is anything, but only God, who makes it grow. The man who plants and the man who waters have one purpose and each will be rewarded according to his own labor. For we are God's fellow workers, you are God's field, God's building. By the grace God has given me, I laid a foundation as an expert builder, and someone else is building on it. But each one should be careful how he builds." 1 Corinthians 3:6-10

The Master Gardener forms a hole in each of us. The emptiness is intentional. Just because we can't see God in that empty place, it doesn't mean He can't see us. Unlike the earth, the hollow place in a man's life can only be filled by the love that formed it. God desires it to be filled by the best He could give. Galatians clearly tells us what that is.

"The promises were spoken to Abraham and to his seed. The Scripture does not say "and too many seeds," meaning many people, but "and to your seed," meaning one person, who is Christ." Galatians 3:16

Life is void without God's filling. What is sown in the vacant places of a heart will breed innumerable wrong choices, unless it is emptied and then filled with kernels that hold God's storehouse of blessings. That is why it is important to know what seed is left in the barn.

". . . Give careful thought to the day when the foundation of the Lord's temple was laid. Give careful thought: Is there yet any seed left in the barn?" Haggai 2:18b

Never releasing into God's hands what we fear to lose does not save it, it holds it back. We are caretakers not owners.

"I tell you the truth, unless a kernel of wheat falls to the ground and dies, it will remain only a single seed. But if it dies, it produces many seeds." John 12:24

What hidden potential is just beneath the surface of a person is brought to fruition by God alone. But God gives us power to water, plant and weed around his gifts. Others will be blessed by them, be drawn to his provision and built up by what we share. What gift and purpose God has bestowed on the most infuriating child, complacent teen, drunken man or desperate woman is not always obvious. What extraordinary dreams God waters in the most ordinary and unremarkable place God plants us will only be unveiled if we garden there.

"He will send you rain for the seed you sow in the ground and the food comes from the ground . . . the ground will be rich and plentiful...."
Isaiah 30:23

me plants, like corn, do best when they are planted large numbers and cover large areas. Other plants, e forsythia bush, can survive alone. Holly bushes and ngko trees cannot produce fruit unless there are both ale and female plants in a garden. However, in public rdens you only want to plant the male Gingko trees. e fruit from the female Gingko tree is very unpleasant. od designed each differently. So people flourish or uggle in different environments.

eds are dispersed by the wind, birds and water. Some ve fancy structures that cling to the coats of animals. me seed coats are softened by stomach acids and e planted in droppings and others have propeller like echanisms that carry them out into the world. In the me way God will close doors in one place and open up portunities in another very intentionally. He sees the ndscape even if we can't see past the soil. God's plan for r lives cannot be thwarted by our willingness.

ow deep a seed is buried helps or hinders its chances survival. Some seeds need to be scattered just under e surface of the soil and others need to be buried deep. hen the seed ceases to take comfort in the darkness d pushes its way out new life germinates. It pushes rough the weight of soil covering it. The darkness at has trapped it loses its hold when it is warmed and awn by power from above. It sends roots deeper, as stem and leaves reach for the light. It grows, moves, aches up and digs in. It begins; that small state of being, anches out as tall as the highest tree with root systems eper than can be easily measured.

"No temptation has seized you except what is common to man. And God is faithful; He will not let you be tempted beyond what you can bear. But when you are tempted, He will also provide a way out so that you can stand up under it." 1 Corinthians 10:13

When God allows darkness to cover us, it is always to bring forth new life and hope. Until we stand, we will not see the heights we can reach, breadth we can reach or depth we can be anchored. But God has a purpose for the darkness as well as the light.

"But this happened that we might not rely on ourselves, but on God, who raises the dead." 1 Corinthians 1:9b

Even after all the time and distance that separates us from Eden; after all the jungles we have planted and forests that have grown between us; despite canyons that led us further from our evening walks with our Creator; after all that has made our lives what they are; He is still waiting to fill us. He knows our lives have unknown potential when they meet the right conditions. Seeds of Eden still sprout hope. They still lie dormant just waiting for us to meet Him in our prayers.

"As long as the earth endures, seedtime and harvest, cold and heat, summer and winter, day and night, will never cease." Genesis 8:2

• Tell about how someone made you feel who will always be remembered. Try to focus on a good memory.

• Has God filled an emptiness in you that you thought could never be filled?

• Have you ever released something into God's hands reluctantly and then watched as it changed into something beautiful?

• Have you seen tough times, darkness or a burial ignite new growth around you?

Waiting...

My kids loved planting seeds when they were little, but quickly were frustrated when after two or three days of checking nothing happened. Waiting is tough. Of course, when nothing came up at all, they were heartbroken. Usually weeks later, they would run in and pull me through the yard to show me that the seeds finally began to grow.

"A farmer went out to sow his seed. As he was scattering the seed, some fell along the path; it was trampled on, and the birds of the air ate it up. Some fell on rock, and when it came up, the plant withered, because they had no moisture. Other seed fell among the thorns, which grew up with it and choked the plants. Still other seed fell on good soil. It came up and yielded a crop, a hundred times more than was sown. Luke 8:4-8 (Luke 8:11-18 the meaning of the parable.)

Seeds lie dormant in a state of quiescence until they begin to grow. They rest before they rise. It is not always apparent what is taking root. Constant careful attention is necessary to keep the weed seeds from taking over planted seeds.

"Sow your seeds in the morning and at evening. Let not your hands be idle. For you do not know which will succeed, whether it is this or that or whether both will do equally as well." Ecclesiastes 11:6

Seeds have diverse structures. Their makeup ensures survival. They can lie dormant for years beneath a farm field and yet crack concrete when they are ready to reach for the sun. Small things can have great power. We know this from Matthew.

"He replied, "Because you have so little faith. I tell you the truth, if you have faith as small as a mustard seed, you can say to this mountain, 'Move from here to there' and it will move. Nothing is impossible for you." Matthew 17:20

A gardener can encourage growth by being aware of germination triggers. Germination triggers help a seed to come out of its shell. In all seeds, the seed coat protecting them from a larger life needs to crack open. "We're all broken. That's how the light gets through." – Ernest Hemingway. Some seeds just need water and the sun. Several seeds do not germinate until they have been soaked. "Life is a storm, my young friend. You will bask in the sunlight one moment, be shattered on the rocks in the next. What makes you a man is what you do when that storm comes." – Alexandre Dumas. Other seeds need to be scratched. Fire alone burns away the protective coating of a few seeds. I wonder if this is why God used a burning bush to wake up Moses. Deuteronomy 33:6. Maybe he took off his shoes to avoid trampling the seedlings. Several need icy temperatures while others can never be in a place where the temperature drops below freezing.

"Whoever watches the wind will not plant; whoever looks at the clouds will not reap. As you do not know the path of the wind, or how the body is formed in a mother's womb, so you do not understand the work of God." Ecclesiastes 11:4-5

God uses germination triggers to encourage new development in us. It's not always easy to trust Him, but we always can. None of us like the fire of stress, the numbness of freezing, the tearing of scratching or the feeling of drowning as He allows our environment to pull triggers in our place of complacency.

"Don't be deceived, my dear brothers. Every good and perfect gift is from above, coming down from the Father of heavenly lights, who does not change like shifting shadows. He chose to give us birth through the word of truth, that we might be a kind of first fruits of all He created." James 2:16-18

Harsh times in our lives are sometimes perfect gifts from above. He uses them to encourage growth in us. He knows what keeps us from growing and what needs to be cracked, softened or burnt away. The hardest things we go through can be triggers for change. Life is desire waiting in boring and stagnant darkness. We are desperate for the light to warm us, water to refresh us, and to take our first breath. God sees what our lives were meant to be and He knows how to trigger growth that has lain dormant. God will allow troubles to bury us and frost to break shells hindering us when it is necessary. Prayer through those tough times will help remind us of what is possible.

"Those who sow in tears will reap with songs of joy. He who goes out weeping, carrying seed to sow, will return with songs of joy, carrying sheaves with him." Psalms 126:5-6

This book was written under such circumstances. I gardened every summer for years. I came to love every moment in a garden. Gardening is my love language and expression of worship for my Creator. It helps me focus and to escape. My body changed and I began to struggle with heat and long hours. I worked through the pain for years, and watched myself age while cooling myself with ice constantly. My doctor warned my heat intolerance would progressively worsen. I had to let gardening as a profession go. I went through a time of dormancy and grief letting go. But in the burial of that passion, God ignited a joy for remembering. Through writing, the shell that encased that love broke. What was planted bloomed in a different way. This book grew in the seedbed of a writer's guild that encouraged its growth. (and I still get to garden in the evenings, when it is raining and any place it is shady and cool.)

"The eyes of all look to you, and you give them their food at the proper time. You open your hand and satisfy their desires." Psalms 145:15-16

Many seeds do not germinate for one reason or another. Germination in the soil can be affected by the slightest chemical imbalance as well as the terrain. What encourages growth in man is affected by their life situation, age, and the troubles that surround them in the time of planting. Our environment and attentions constantly shift. Sometimes we do not know how to encourage growth in another. Similar age or life situation can be helpful. Often we need to share a common life experience or be focused on the same things to be able to encourage another's growth. Other times until we are ready to receive the seeds God wants to plant, they cannot grow in us.

"Remember this: Whoever sows sparingly will also reap sparingly, and whoever sows generously will also reap generously."
2 Corinthians 9:6

New seeds can always be replanted. In life, few lessons are learned after the first tutorial. Most inventions are only discovered after many failures. Not everything that can grow in us has all the ideal conditions met in the same moment. In the abundance of the Lord and a garden, plant and sow generously. Be willing to wait for what the Lord is growing.

"I wait for the Lord, my soul waits, and in His word I put my hope." *Psalms 130:6*

• Why is waiting so hard for us? Why is waiting important?

• Do people have germination triggers? Give some examples.

• What helped you come out of your shell?

• What is it about a storm and saturation that wakes us up?

• How is a destructive fire a different awakening?

Consumption...

Mom's memories shaped my thought life as a child. She talked mostly about things she loved. Conversely, some people dwell on memories that hurt and haunt them. Unfortunately their lives are formed by what manifest from such reminiscences. Misidentification of seeds destroys beautiful places and underestimating the power of words corrupts innocent hearts. The seed we disperse in our children's dreams have the potential to produce a hundred fold in their futures. Passion and hurry continually plow soil and drop unintentional seeds if we are not careful to listen to God's voice.

"Then God said, 'I give you every seed-bearing plant on the face of the whole earth, and every tree that has fruit with seed in it. They will be yours for food.'" Genesis 1:29

We are free to choose what we want to plant in any gardens and in any life. It does not mean we always make the right choices. Good gardeners take into consideration that he consumes what he harvests from what he plants. The consumption of it ultimately defines his health.

"See what the land is like and whether the people who live there are strong or weak, few or many. What kind of land do they live in? Is it good or bad? What kind of towns do they live in? Are they unwalled or fortified? How is the soil? Is it fertile or poor? Are there trees there or not? Do your best to bring some of the fruit of the land?" Numbers 13:20

Most harvested seed is used to fill our bellies. Nuts, especially, are packed with good things, have a long storage life and can be a staple for survival.

I gardened for a woman in her 80s. She worked beside me every day, and she wore me out in my 30s. She continually claimed her secret was 12 nuts a day. I believe her. Seeds are storehouses of energy. When digested nuts and seeds have naturally occurring magnesium. When we have enough magnesium in our system, it influences the production of serotonin in our bodies which helps us to feel good and have a positive mindset. Seeds along with flowers, for grieving people, may be a good way to help them to get through their time of loss.

The seeds we use to feed ourselves, we also feed our children. Though many are tiny, some are large enough to choke a person. The largest seed is the coconut. Some like the avocado seed, if you succeed in swallowing it, might kill you. There are also poisonous seeds. Stone fruit seeds come from peaches apricot and cherries. Though I have read that in low doses they can be consumed for health benefits. So educate yourself. The wrong seeds scattered in children's lives or fed to them in times of desperation ultimately can cause them more harm than good. But seeds of good things imbue strength, wealth, generosity and thanksgiving.

"Now He who supplies seed to the sower and bread for food will also supply and increase your store of seed and will enlarge the harvest of your righteousness. You will be made rich in every way so that you can be generous on every occasion, and through us your generosity will result in thanksgiving." 2 Corinthians 9:10-11

Fruit left on the vine rots, along with it, the seeds it harbors. Saving and storing fruit and seed must be intentional. The seeds of one or two tomatoes need to be saved for every 100 or more consumed in summer.

"The seeds are shriveled beneath the clods. The storehouses are in ruins, the granaries have been broken down, for the grain has dried up." Joel 2:17

Learning to save seed can be a life saving lesson. In a Beth Moore Bible study, she shared a story of seed brought to villagers in a desolate land. The people who sent it intended the seed to be used for planting. But the villagers were so hungry; they ate the seeds in desperation and ignorance. Man's inability to understand what he has been given and the power it has to impact his and other's lives ruins storehouses of blessing.

"Their destiny is destruction, their god is their stomach, and their glory is their shame." Phillipians 3:17

Initially we are all consumers. As children we need to be fed to survive. But God is calling us to be producers and planters. We were made to be gardeners. He wants us to grow up and harvest seeds to share. The lack of knowledge causes treasure to be expendable and trouble to fill our empty appetites.

"My children perish for lack of knowledge." Hosea 4:6

• Were there seeds planted in your childhood you saw come to fruition in your adult life?

• What are examples of seeds we save in life to plant or share in the future?

• Can you think of an example of seeds consumed to fill an appetite that were meant to be planet for a harvest?

Harvest...

"I am the vine, you are the branches... if a man remains in Me and I in him, he will bear much fruit; apart from Me you can do nothing." John 15:5

I remember going for walks with my grandparents along gardens and paths encircling their property. I do not remember a word my grandparents spoke. I was very little, but I planted memories on those walks. They were scattered and sprouted along the paths I walked throughout my life. I gathered harvests from them. They fed me. I shared fruit and seeds from them. Those seeds rooted in different places, but they are united in source.

"I tell you, open your eyes and look at the fields! They are ripe for harvest. Even now the reaper draws his wages... 'One sows and the other reaps,'... I sent you to reap what you have not worked for. Others have done the hard work, and you have reaped the benefits..." John 4:35b-37

No man consumes every blueberry grown by just a single blueberry bush he plants in His life. But it is possible that others will gather the fruit he planted decades after him. We see, cut and enjoy but a tiny portion of the flowers that bloom from the flora seeds we plant.

Our own hands do not always get to pick the fruit we planted the seeds of. Often times it is harvested by others. Ralph Waldo Emerson said, "The earth laughs in flowers." Is the gardener still sharing pieces of himself, the gift of his life through the laugher he planted in the garden? We will not always know the impact our lives make. But what we plant always goes on to feed others, whether it is fruit for the eye, soul or body.

Harvest is the season when ripened crops are gathered, a crop or yield of one growing season. Fruit is any part of a plant useful to humans or animals; the developed ovary of a seed plant with its contents and accessory parts.

Everything about us, from what our mouths utter, to our passions, pursuits, and darkest fears plants seeds. Our habits, wasted energies and disappointments plant seeds as well.

"Now this is what the Lord Almighty says, "Give careful thought to your ways. You have planted much, but have harvested little. You eat, but never have enough. You drink, but never have your fill. . ." Haggai 1:5a

Each seed is precisely what it is made to be. It is a womb waiting with hidden and underestimated potential. We need to plant that which produces worthwhile things for all the effort of harvest. Our choices are so important because something always grows from them.

". . . My brothers, can a fig tree bear olives, or a grapevine bear figs? Neither can a salt spring produce fresh water." James 3:11-12

Poison ivy does not produce blueberries. But many animals of the forest eat it. Many deadly plants are used in medicines. Cherry trees will never be able to bear bananas. That is not their purpose, nor their programming. It does not make them bad. It makes them specific.

I hope my kids read this someday and understand why so many of the programs they wanted to watch on television were off limits. They planted nothing worth caring for and neglected to provide anything of value.

A new place is created when one (purposely-planted) seed at a time is added to a garden. There is possibility, new chance, and opportunity in each.

Harvest is the purpose of our lives. Plants know this. If a plant is struggling, it will push one last time to flower and produce seed even at the expense of its life. I will often cut the flowers off a struggling plant or newly planted shrub. This will keep it from going to seed. It will then put its energy into its roots systems. I know that God has cut away what I thought was beautiful about me so that I too would dig a little deeper and have more to depend on than appearance.

My sister-in-law's crabapple was stripped of bark one winter by rabbits. In the spring before dying it bloomed more profusely than ever. Some of the most beautiful memories of a life come to fruition when a person leaves a blessing even in times of heartbreak. When we fail to bloom, there will be no harvest. If we fail to be productive we will depend upon the fruit and beauty of others. Let us not grow weary of planting blessings for the future and in our children's hearts. Our time is short. Planting seasons and seasons of harvest do not last forever.

"The harvest is past, the summer has ended and we are not saved."
Jeremiah 8:20

It is not what we hide in storehouses that make us rich. The Bible says,

". . . a man's life does not consist in the abundance of his possessions."
Luke 12:15b.

What a life harvests and shares is what defines it fruitfulness or futility. What seeds, memories, and experiences are cultivated in it promote its fruit. It is what we never stop believing, cultivating and offering that produces; love, joy, peace, patience, kindness, goodness, gentleness, and self-control. (Galatians 55:22)

"Joy is not a feeling but a FOCUS." - unknown author

How fruit is harvested will vary. Grapes sweeten more if they are watered less right before harvest. Southern exposure also helps them produce a better harvest. Strawberries are sweeter if it frosts right before they are harvested, but only if their skin is not damaged. Peaches, tomatoes, apples and grapes produce better crops if their plants are pruned in a particular way. My sister-in-law can grow a 10-foot tall tomato plant on a single stem and it is full of tomatoes. Most fruit ripens best if it is grown in full sun. Mushrooms, of course, fruit best in the dark. Sometimes God uses a dark time to bring forth fruit in us. Planting may not always ensure harvest, but it is the only hope we have for one. All fruit sweetens best if it ripens on the vine or tree.

"No branch can bear fruit by itself; it must remain in the vine. Neither can you bear fruit unless you remain in me. I am the vine, you are the branches... if a man remains in me and I in him, he will bear much fruit; apart from me you can do nothing." John 15:3b-5

We are all branches that need the body of Christ to make us capable of bearing fruit of value. We need the support of the vine to keep us from being trampled by the world. The Master Gardener encourages us to fruit and the banquet table is overflowing with variety. Most of the world is content to sit still in purposeless activities asleep in comfort or isolated hollow buildings. "Where you invest your love you invest your life." – Mumford and Sons. Fill those voids with sweet fruit, fellowship and firm foundations. Spend time commemorating and building. Plant much with the hope of harvesting much.

A life is used up working for things that are never enough, never filling and endlessly insufficient. We all go to seed. There are seedy characteristics about each of us. If gardened, a life can be quite beautiful despite all the waste.

"But I said, "I have labored to no purpose; I have spent my strength in vain and for nothing. Yet what is due me is in the Lord's hand, and my reward is with my God." Isaiah 49:4

The seeds we plant are the legacy we leave for tomorrow. The memoirs we share keep things alive that would otherwise disappear. Even the smallest seeds we share can capture a tremendous amount of His light. Great things start as small things. Fruits and blessing reach over boundaries, cascade beauty and nurture others if we plant them in God's care and harvest them in their own time.

"The land you are entering to take over is not like Egypt, from which you have come, where you planted your seed and irrigated it by foot as in as in a vegetable garden. But the land you are crossing the Jordan to take possession of is a land of mountains and valleys that drink rain from heaven. It is a land the Lord God cares for: the eyes of the Lord your God are continually on it from the beginning of the year to its end." Deuteronomy 11:10-12

- Do you remember walks as a child? Are there flowers that change your mood?

- Have you seen a damaged tree flower in excess before it died? A damaged soul or life ending that bloomed?

- Does being a part of the body of Christ make you feel like you've been seated at a banquet table? or like visiting a hollow house?

- Do you know a rich man whose wealth has nothing to do with his possessions?

GARDEN
Always

Concerning Weeds...

"To the weeds. To the wilderness." Mary Reynolds in Dare to be Wild

Weeds have a bad reputation. Weeds aren't always a bad thing. In most of the world, their wild foliage anchors the soil in place. They stabilize the environment. In the movie Dare to be Wild, Mary Reynolds encourages the world to preserve the wilderness. At times, we all need the wild places of the world as much as a guarded garden. The best way to preserve them is to leave them untouched.

I spent huge amounts of energy grumbling about weeding the garden at my parent's old farm house. Mom had a large garden behind the barn. Anytime the volume in the house needed turned down, she sent my two brothers and me out to weed. I thought it was mom's way of torturing us. We were closer in ages than my three sisters, so we did most things together and usually as a unit fought against them. I am sure mom used distance to lessen the fighting. When separated from our enemies, of course, our battles were redirected at each other.

From one day to the next, I never knew if my hair would be pulled out, if I would be sprayed with the hose or have porcupine eggs thrown at me. "Porcupine eggs" were our secret weapons. Right beside our garden, a Chestnut tree dropped nuts wrapped in a spiny coats, hence their nickname. Even if they were gingerly picked up, they sent spikes of pain through fingers. So we carried them carefully and tossed them to each other frequently. Hollered "catch!" and ran. We would plant them in each other's shoes and a few greeted us in our beds. Hence, I spent a lot of time attempting to predict my brothers' next moves and trying to catch them off guard.

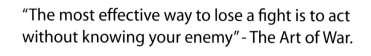

"The most effective way to lose a fight is to act without knowing your enemy" - The Art of War.

Brothers and sisters know their enemies quite well, and the battle is never over. Now that we are adults and appreciate each other so much more, I realize that I was fighting the wrong battles back then. Since that time, I have been trying to identify the difference between my **ene**mies and the weeds I misidentify. Then **I take time to know** them before I take **action against them.**

When we traveled out west and visited Yellowstone and Glacier National Park, I realized that the wild flowers grew like weeds. They were beautiful just as they grew. I remember how much my brothers and I fought. I think

about the adventures we had together. They are my favorite memories. We would wander into the forest or down to our pond and get lost in time.

There are many **different types of weeds in this world. They have artful means of dispersal and can lie in wait for just the right time to germinate. Some grow ominous forests clouding out the light. The manifestations of** others writhe across the ground just a few inches high, tripping us like little white lies. Some are ugly all the time like hatred, and others, like lust, mimic beauty but secrete poison. In a garden weeds need to be identified and confronted. Learning to establish battle lines for the skirmish is crucial for life. If we fear the thorns, we may fail to meet the flowers.

Solomon said, "All things are wearisome, more than one can say." Ecclesiastes 1:8a. I imagine he said it one morning as he walked out through his gardens and noticed the weeds erect and ready for battle. Weeds are relentless.

The world is a crazy place. Even a garden can be confusing. The weed one gardener pulls, someone else may plant. For example, crown vetch is planted along the highways here in Pennsylvania. It holds sloped hills in place with deep root systems and can be attractive. In a garden, crown vetch is a scourge. Its roots are deep and the plant surfaces everywhere. Early in spring, dandelion leaves are a nice addition in a salad. But its little yellow head and whispering seed are considered an archenemy by most that battle it in lawns.

Weeds are carefree sparks of chaos, whether blown by the wind or carelessly dropped by a gardener. They have qualities that are more detrimental than beneficial. Some have minor shortcomings, others gigantic flaws. They spit seeds where they are not wanted and send out tentacle-like roots that wrap around diminish purposeful plantings. They may be a normal part of a garden since the fall, but by no means are they wanted members.

In unplowed ground, weed seeds can lie dormant for years. Any attempt to cultivate the soil provides oxygen and a glimpse of the sun; the two things needed for germination and so weeds pulled day after day are but a fragment of the reinforcements that surface year after year. Not all of them produce thorns, but in a garden they steal space and beauty.

In Hosea 10:8 NLT, we learn, "... the place of Israel's sin, will crumble. Thorns and thistles will grow up around them..."

Often sin is camouflaged like a vine that slowly grows up the center of a tree and then out over its canopy. It can flourish when it is not anticipated. Sometimes is like poison ivy. Fear keeps us from confronting. Weeding then, as it parallels life, is identifying undesirable places where thorns and thistles have a foothold.

"... Every plant that my Heavenly Father has not planted (needs to) be pulled up by the roots..." Matthew 15:13a,

Gardens have perimeters. Weeding outside of that assures us of failure. When we weed our lives correctly, we purposefully remove all that hinders a relationship with our Creator. As in Hebrews 12:15, "See to it that no one misses the grace of God and that no bitter root grows up to cause trouble for many." To weed a life is to pull up by the roots, anything that blocks the Lord's light or hinders showers of His affection.

• Have you ever lost a fight because you underestimated or didn't know your enemy?

• Which weed do you battle most often in your garden? Is some part of it attractive? What do you think makes it a weed?

• Is there anything flourishing in your life that blocks the Lord's light and hinders showers of His affection?

Recognition...

My kids are very allergic to poison ivy. So I point it out to them every time we see it. It is not always easy to identify. The leaves are always similar, but this plant takes different forms, depending where it grows. My kid's best defense against it is an ability to identify it. Yet my son came home from playing paintball with friends covered in a poison ivy rash. He said he thought he would be OK. He was having a good time. He just laid down in it for a little while.

The physical characteristics of some plants do not reveal the aggressive qualities of their nature. The cost of some choices is not always obvious at the moment of decision. So life can be plagued with sin that seems harmless in any given moment.

Sometimes we are unsure of what is growing, and we wait to weed. Seedlings start off small. They are not threatening. We often do not fear them until they grow big enough to reveal what is hurtful about them. By then often, they are entrenched and a battle must ensue. In The Little Prince, the baobab was something that could never be gotten rid of if it wasn't dealt with soon enough. A planet was torn apart when they grew too big to control.

Life teaches us the same lessons. If unrecognized evil things can multiply unchecked. If unrestrained, even good things can diminish us. For example, eating and drinking can be very pleasing and healthy activities. Yet gluttony and drunkenness grow from mutated forms of them. Hurt feelings can put down roots and produce bitterness. Worry that is not checked grows into crippling fear. Loneliness strangles; insecurity flourishes into vulnerability. For all the wrong reasons, jealousy and envy develop into murder and stealing. Arrogance and pride defeat us with great ease because of silent and destructive powers. The list continues.

"The earth dries up and withers, the world languishes and withers. . . The earth is defiled by its people; they have disobeyed the laws, violated the statutes and broken the everlasting covenant. Therefore a curse consumes the earth. . ." Isaiah 24:4-6

Understanding the intimate qualities of small things taking root around us and in our hearts is what weeding is about. The opportunity to pull sin away in its inconsequential state is lost in poor judgment.

"I went past the field of the sluggard, past the vineyard of the man who lacks judgment; thorns had come up everywhere, the ground was covered with weeds, and the stone wall was in ruins." Proverbs 24:30-32?

Recognizing under what conditions weeds are established will help us to perceive when extra care may prevent them. My neighbor is a farmer; he told me that weeds are indicators. Particular weeds can be established because of the predisposition of the soil. Those weeds tell us what could haunt a place in the future. Some soils have properties that encourage the excessive growth of typically harmless plants that then become weeds. Many people can work in and around poison ivy. It has not effect on them. In the same way the makeup of some people's soil can cause them to struggle with weeds or sin that another's constitution easily has power over. Each of us must,

"Be careful . . . that the exercise of your freedom (the weeds we are able to control easily) does not become a stumbling block (or source of weed seeds) to the weak."
1 Corinthians 8:9

At times our actions cultivate weeds we have planted ourselves, but at other times weed seeds collect in plowed soil dropped carelessly by the wind into an open wound. Others are purposely pushed into wounds by an enemy taking advantage of grief, exhaustion and desperation.

Jesus told them another parable: "The kingdom of heaven is like a man who sowed good seed in his field. But while everyone was sleeping, his enemy came and sowed weeds among the wheat, and went away. When the wheat sprouted and formed heads, then the weeds also appeared. The owner's servants came to him and said, 'Sir, didn't you sow good seeds in your field? Where did the weeds come from?' 'An enemy did this.'" He replied. Matthew 13:24-32, (explained 13:36-43)

A friend shared the story of a time of grief. She lost her husband. Her world was plowed under and she had no desire to get out of bed. Her home and garden were neglected and weeds grew in her sorrow.

". . . Do not be afraid, though briers and thorns are all around you. . ."
Ezekiel 2:6a

Spring returned in her life, and she returned to her garden. But she had to replant a few of her favorite plants because the weeds took advantage.

The world is a confusing ride of up and down emotions, not to mention right and wrong decisions. "The hungry consume his harvest, taking it even from among the thorns …" Job 5:5a. It is full of people starving for the truth and nauseous from spinning on it axis.

We need to solicit understanding as the servant in Matthew 13:27-30. Recognition will help us to spend more time nurturing what we love and less battling what we don't understand.

"A wise man has great power, and a man of knowledge increases strength; for waging war you need guidance, and for victory many advisers." Proverbs 24:5-6

• Can you share a time when you were unsure of something and allowed it to grow, then faced a difficult battle?

• Have you seen sin take root in an open wound?

• How has guidance and advise changed the balance of power around you?

• What do you do when you find yourself in the midst of weeds you never planted?

Worry Weeds...

Worry can seem like a small thing. It can even be worn as a badge of honor. I have uses it to get information by saying, "I am worried about so and so." Whether I mean it to or not it usually leads to gossip.

"Still others, like seed sown among thorns, hear the word; but the worries of this life . . . come in and choke the word, making it unfruitful." Mark 4:18-19

Worry wastes energies on things that do nothing to produce fruit and blossoms in life. Fruit and beauty come from getting on life's knees and pulling weeds. It needs plowed into the dirt and cultivated in prayer. When doing anything else is beyond our power, action is the only thing that amends it. Worry doesn't stop death. It stops life.

Worry is a sin, because it replaces God with something as insignificant as a weed. It diminishes our perception of His power, extinguishes joy, and saps strength. He alone has the power to weed things from life that are not under our authority or control.

Other's opinions are often a source of worry. At a time when I needed to hear it, someone told me, "Some peoples' opinions do not matter." Nothing holds us back from fruition like the fear of judgment or mockery of those we allow to stand over us. And those who se position to intimidate often have a crowd that cowers around them. This verse helped me put that worry to rest.

"Like a thorn bush in a drunkard's hand is a proverb in the mouth of a fool." Proverbs 26:9

God never treats us like that. Sin is any distraction, including a person's judgment, which defeats the Lord's purpose in us.

Consider this Biblical message:

A thistle . . . sent a message to a cedar . . . "Give your daughter to my son in marriage." Then a wild beast . . . came along and trampled the thistle under foot. You say to yourself that you have defeated (another), and now you are arrogant and proud. But stay at home! Why ask for trouble and cause your own downfall and that of (others) also." 2 Chronicles 25:18-19

It can be hard to remember an evergreen truth when we are small and sin looms over us. Cedars are evergreens. They take years to grow to a height of six feet. The thistles here in Pennsylvania easily grow to six feet in three months. But if given time, a cedar towers over the thistles year round for decades once established. But its growth takes a very long time. As God's people, we are cedars with great potential. We should not consent to unions with the thistles of the world. Our character and our family's repute are married to those reputations. Thistles' lives are but a season. God's purpose for our lives towers over the weed's ugly existence, if we reach for the Son. We must remember whose we are and what we will become under His care.

"Cast all your anxiety on Him because He cares for you." 1 Peter 5:7

Jesus and the disciples walked past a fig tree with no fruit. A fig tree is unlike most fruiting trees in that it produces fruit year round not just seasonally. Jesus cursed the tree and walked away from it. "In the morning, as they went along, they saw the fig tree withered from the roots. Peter remembered, and said to Jesus, 'Rabbi, Look! The fig tree you cursed has withered!' 'Have faith in God,' Jesus answered …'Therefore, I tell you whatever you ask for in prayer, believe that you have received it and it will be yours …'" Mark 11:20-21

In this Miracle of Death, I see a weed that had the appearance of something fruitful, but its promise was empty. I see a God whose words and will had no trouble overcoming it.

Today, it is a small thing to cut down a tree. We have chainsaws. In Christ's day a tree rotted in its place until it fell or someone spent hours at it with an ax. What we fail to observe in the miracle of death is the silhouette of a tree on the horizon for years after its demise. The tree's life and source of power had failed, though it stood in plain sight tall and erect. When God withers the roots of enemies, they are conquered, though they stand for years over us in their defeated states.

Worries, big or small, only have the power we give them over time and happiness. The tree in these verses bore no fruit, nor offered shelter or shade for those who needed it. Big as it was, no moment of worry or diminished happiness would restore it.

We recognize God's gifts by their fruit. "By their fruits you will recognize them. Do people pick grapes from thorn bushes, or figs from thistles?" Matthew 7:16. When worry does not hold power over us, God takes His rightful place. He needs only a word to subdue any obstacle in life's garden. The fruit from His hand is always sweet and perfect like strawberries, blueberries, peaches, raspberries … There is no foul aftertaste and no regrets attached.

• What weeds do you worry about more than God's approval?

• Have you ever seen the Lord remove the power of an ominous enemy though it had the appearance of strength?

• What helps us best differentiate fruit from God's hands and poison fruit?

• Have you ever seen a cedar finally stand above thistles that had crowded it at one time?

A Garden Observed: CULTIVATING A LIFE

Concerning Wealth and Other Things...

"Still others, like seed sown among thorns, hear the word; . . . the deceitfulness of wealth and the desires for other things come in and choke the word, making it unfruitful."
Mark 4:18-19

As for the deceitfulness of wealth, I note the story of Jack and the Beanstalk. Some weeds, like Jack's beanstalk, grow quickly. They are planted in desperation and grow weeds that invite unknown perils and arouse evil desires. Jack quickly became a thief and then a murderer. Great wealth is built over the ruined fields and lives of men who misidentified financial gain (or stability) for wealth.

Wealth can be confusing. We have all watched men and women throw away priceless God-given gifts in pursuit of wealth. The kings of Israel who pursued what they wanted and served their own purposes were defined by their choices. Each lost the Lord's blessing. Temptations that direct us away from God's will can tear apart garden kingdoms as weeds devastate a garden.

Greed destroyed my ability to enjoy Christmas shopping years ago. I wanted something I could not afford so I made myself miserable over it. I was so thankful when God revealed the root of my frustration. It was greed, and then I let it go. Such things are a curse and fire is the only thing that subdues them. I anticipated a laborious task sorting through the desire for other things mentioned in Mark 4, but truth is simple. It is summed up in the wisdom to identify the difference between what is wanted and what is needed. Each man's want list can be extensive and depending on his place in society and age can vary greatly. But, God gives us all very similar needs.

Not all wants are weeds. However, want lists can be huge and our contemplations of them torment if greed takes root. Any want that takes our focus off the Lord is a weed.

The following verse shows us how the Lord's temple changed. A precious place disappeared into a deserted territory.

"In that day, in every place where there were a thousand vines worth a thousand silver shekels, there will be only briers and thorns . . . As for the hills once cultivated by the hoe, you will no longer go there for fear of the briers and thorns; . . ." Isaiah 7:23, 25

Weeding is dirty work. Flinging dirt hitting you in the face as roots relent is drudgery. And don't even get me started on gnats that dive for your eyes. However, the pain that comes with weeding doesn't eat away at your soul like greed can. There is purpose in the pain. It restores the garden and somehow the gardener at the same time. We never had extra money at the farm, but I often went to bed feeling like I had great wealth. The garden was tidy and fruit was plentiful.

"But land that produces thorns and thistles is worthless and is in danger of being cursed. In the end it will be burned." Hebrews 6:8

Weeding always takes patience. It cannot be done well if it is done quickly. In Pennsylvania, there is a vine we call tangle vine or binding weed. Both names are fitting. Each vine needs to be followed to the root which needs to be dug out carefully. The root has runners anchored everywhere. They also need to be followed beneath the soil bit by bit. Watching the perennial's base for any returning vines is a long-term commitment.

When a deep root is pulled it brings to the topsoil micronutrients that can enrich the soil. God can use even the roots of evil for His good when they are dealt with. When we confront what is wrong, God can surprise us with what He plants in its stead as in James.

"Therefore, get rid of all moral filth and the evil that is so prevalent and humbly accept the word planted in you, which can save you." James 1:21

The desire for other things can take God's people's focus off gifts God had already given them. No place is a sanctuary where God is not treasured more than other things and wealth. Without His blessing we wear ourselves out and gain nothing.

"They will sow wheat but reap thorns; they will wear themselves out but gain nothing. So bear the shame of your harvest because of the Lord's fierce anger." Jeremiah 12:13

- Has a desire for other things ever robbed you of joy?
- How has God helped you to recognize its source?
- How was your joy restored?

A Time To Refrain...

Any parent has a hard time watching their children make mistakes. We know what can be avoided if we step in and help. However, it is OK to let our kids make their own choices and struggle with the results even when we see a detrimental outcome. If it is hard to watch our children struggle, it is torment to watch grown adults scatter their own weed seeds purposefully with no vision of consequences.

Some of life's most valuable lessons are learned when weeds overtake what was once was dependable.

"The kingdom of heaven is like a man who sowed good seed in his field. . . then the weeds also appeared. . . The servants asked (the land owner) him, 'Do you want us to go and pull them up?' 'No,' He answered them. 'Because while you are pulling weeds, you may root up the wheat with them. Let them both grow together until harvest. . ."
Matthew 13:24-32, (explained 13:36-43)

It is hard to watch weeds blanket a territory when we know we could, at least, try to pull them. However, weeding can rape the land. Our clumsy bodies break stems and dragging weeds away can tear off buds. Waiting is sometimes the only way to save a harvest. ...there is "a time to embrace and a time to refrain ..." Ecclesiastes 3:5b

The story of Lot illustrates a time when God allowed a man to be established in a place overrun with weeds though it had the appearance of a lush garden. 2 Peter says,

"...and if He rescued Lot, a righteous man, who was distressed by the filthy lives of lawless men (for that righteous man, living among them day after day, was tormented in his righteous soul by the lawless deeds he saw and heard) - if this is so, then the Lord knows how to rescue godly men from trials and to hold the unrighteous for the day of Judgment, while continuing their judgment." 2 Peter 2:7-9

Lot built in a place that appeared to be a good choice. So God allowed him to be established in the mist of weeds and did not intervene in his decisions. In the same way if we weed too aggressively around our children and do not allow them to see the ugly mess weeds make, they could grow to resent us and learn nothing from our help. Sometimes a person cannot appreciate true beauty until he or she walks out of the weeds, jungles, and ungoverned consequences that form a place unattended.

Mom refrained from weeding the gardens at the farm. She had us do it. She taught us what we needed to know and trusted us to make our own choices about the weeds. When we weeded, she remembered to bring out a snack or cool drink. Dad would sometimes show up, pick up the weed piles and carry them away as well. The extra help made a big difference. The time it takes

to refresh those struggling through daily life, plants beautiful things in the soil of relationships. When we take time to lighten each other's burdens, it will not easily be forgotten. The parts of the harvest that could be lost without help are saved, while our lives become a cool drink of water to those surviving the weeds of the world.

I think when God met Adam each evening in Eden Genesis 3:8, God meant for the visits to focus on Adam's thoughts about his life and the world around him. The Creator of the world was building a relationship. He may have wanted Adam to know how and for what reason He made the world for him. But I also think, He wanted Adam to know his heart and to be formed by His thoughts. Weeds look like weeds when our eyes are focused on what is truly beautiful. Without even knowing it, our hearts and lives are refreshed while we weed.

My kids are always waiting for me while I hurry around my home or rush out the door in pursuit of lesser things. They need me to enjoy their hearts and minds, before I tackle their weeds. Sometimes we go for walks and I remember the adventures my brothers and I had. We weren't worried about troubles that were not ours to pull. A gardener needs to remember to let the garden fall away. Focusing on the faces and hearts of those who need to be refreshed and escape the world, if only for a moment, clears away more weeds than anything. Visits form the hearts of children, heal the broken and embrace the lonely. As it was important for Adam to spend with the Lord each day, it is important for gardeners today to meet with Him. We can then take the part of God He shares to others. In our time with Him, we learn to distinguish His gifts from the weeds. We learn to appreciate the wilderness, but not get lost in it.

Have you ever watched someone walk out of a mess and seen them appreciate the gardened territory because of it?

How hard is it to let our children make their own choices and watch them be disappointed?

How has your alone time with the Lord shaped your choices?

Humility...

I love gardening. At home I am never happier than when I am playing in my gardens. Yet as I weed I try to point out my weaknesses to God. I tell him about what I am struggling while I weed and thank him for blessings as I water.

However, when I first started working as a gardener I had a hard time kneeling in "foreign soil" a.k.a. other people's homes. Going to work in dingy clothes without makeup or pretense was out of character. I fought a feeling of humiliation. I did not like being a servant, even when I was paid to do what I loved. I look back now and realize pride and insecurity confined me.

"If you are wise and understand God's ways, prove it by living an honorable life, doing good works with the humility that comes from wisdom."
James 3:13

Unfortunately, this was not the only evidence of my pride. I have a sister seven years younger and in my adolescent mind, she was inferior. I never played with her willingly and often disciplined the things about her I had just outgrown myself. As adults now, I realize that pride kept me from being a good big sister. I did not garden for her. Instead I threw my weeds into the patch of soil from which she was trying to draw beauty. I wish I could go back and amend my misgivings. Humility and kindness would have made our relationship a blessing. It breaks my heart to see her cry as an adult when she tells me she thought I did not like her when she was little. My treatment of her formed her opinion of herself and her relationships. She lacked the friend and ally I might have been if i had recognized this sooner.

"Humility and the fear of the Lord bring wealth and honor and life." Proverbs 22:4

An insecure, prideful person will not wrap a cloak of humility over their shoulders without a fight. It is hard to nurture beauty in and around others when we do not see it in ourselves. A curse was born the moment pride positioned us in the place of God. It started our problems and continues to fuel them. Where pride raises its head, we fail to look down. Weeds creep up. While we refuse to kneel and pull them, they send unseen runners deep beneath the surface. They grow and multiply whether we recognize their presence or not.

" . . . First (recognize the weeds in and weed your own garden), and then you will see clear enough to deal with (see the weeds) in your friend's (garden)." Matthew 7:5

I later worked with people who made me laugh and told light-hearted stories. I never noticed their haggard clothes or lack of makeup. I was never self-conscious about mine either. I learned from them that gardening is about encouraging beauty in the world, (in others) not about obsessing over my own. A person with a servant's heart can kneel in another's garden and pay attention to its needs. We all have ugly growths, thorns, and blemishes we wish to hide. There are things about ourselves we cannot change, but there are things in the world we can when we are willing to humble ourselves and consider the needs of the world as our own.

"Do nothing out of selfish ambition or vain conceit, but in humility consider others better than yourselves. Each of you should look not only to your own interests, but also to the interests of others." Philippians 2:3-4

Pretty places provide no evidence of pain that goes into their creations. On our knees, we each have a clearer vantage point. Rough hands are familiar with every inch of soil and stem. Children and those bent over by the weight of the world can approach us on their level. Fancy clothes and makeup covering our fault will never be able invite the friendships that a smile and a bouquet of flowers do with so much grace.

Gardening is intimate. It requires eye contact and hands-on relations. Put on something that can get dirty, take off the makeup and initiate negotiations on your knees. The Bible tells us to do this.

" . . . All of you, clothe yourselves with humility toward one another, because 'God opposes the proud but gives grace to the humble.'" 1 Peter 5:5

Beauty exists in the world even if we cannot see it, because He left it here for us. When people around us are exhausted by their burdens, the time is right to offer to help with the weeds. When He is the object of our greatest love, our works and passions cultivate His kingdom. Even if we are covered in dirt, our lives can be a favorite focal points in the world's gardens.

"The worth and excellence of a soul is to be measured by the object of its greatest love, of which its works and passions depend." - Henry Scougel

• What is the object of your greatest love? What fuels your works and passions?

• Do you have trouble taking off your mask and wearing work clothes around others?

• Challenge yourself. Bring a bouquet and snack to a friend's home this summer then stick around and help weed the garden. You will be surprised what God can grow.

GARDEN *Always*

Concerning Compost...

Shadowlands is a movie about C.S. Lewis's brief marriage to the love of his life. He had finally come to the place where he didn't. "want to be somewhere else anymore. He wasn't looking for anything new to happen. Not looking around the next corner, over the next hill." It was his kind of happy. They only had a couple years together before she died of cancer. In one scene she forces Lewis to talk about her coming death. She loved him enough to confront the truth that the pain of death is a part of the happiness now. This truth didn't spoil happiness. It made it real. Much like a fire death changes everything.

Exploring Glacier National Park this last summer we hiked through an area devastated by forest fire a few years earlier. The burnt out part of the park had a powerful waterfall pushing a river through it. Energy and oxygen infused the atmosphere around it. The towering charred tree trunks still stood though ghostly gray. The vegetation was returning, but fire changed all that existed before it came. In truth, even national parks protected by our government and countless people who love them come to harm. However, even fire couldn't change what the place could become. So the land grew into what it was while we visited. I'm sure it felt like a different place with trees towering green and lush over it and will again in years to come be different again. But the place still felt like magic even with dead trees standing century over it.

On a less tragic scale. One spring, I mowed down my mom's peonies as they were beginning to sprout. They were my favorite flowers. I pined for them all summer. To my amazement the next spring they grew again. God's creations are amazing things. It doesn't change the fact that I was supposed to be caring for them, but instead harmed them. Many things in the garden have gifts that regrow. If foliage is frost bitten or a dog tramples your blooming bulbs, it is OK. Perennials have all the tools they need to grow again each spring and many current-year leaves and stems will grow again if the browned and broken foliage is cut away.

If border plants are taller than expected and background perennials are dwarfed by mid-ground shrubs, it is OK. Gardens can be rearranged fairly easily. Most, though not all, plants can be transplanted. Seeds are inexpensive. Divisions are free. Dividing roots can be a messy job, but I wish fresh starts and do-overs were so easy in life.

There are a lot of things in my life I wish I could change. In adolescence, we are wise fools growing in both size and ignorance. I made more hurtful choices than I made friends. It is hard to be a target, but I think being overlooked hurts just as bad. Meeting rejection for the first time myself, I wish I had not added to other's wounds. It is amazing that some of those injuries are still tormenting. This awkward time cannot be avoided. Adolescence is an accumulation of clumsy and hurtful things. It is a part of life's compost.

Life changes with the power of a waterfall. One minute the sun is sparkling off a gentle stream. Turn around and the water's power washes your feet out from under you. One visit you're a child amazed, the next you watch your kid tumble in and your heart sinks beneath the water with them. Every season everything changes. The water's flow never quits. It's too fast, too strong, overwhelming; it's never enough.

"For small creatures such as we the Vastness is bearable only through love." - Carl Sagan

For all the wonderful things about this world, it is also compiled of what we lost, neglected, wounded, and even killed. What we want to be is not always what we are. The same is true for a garden, what we dream of creating and end up growing are often very different realities. It cannot be avoided. Those realities are a part of what forms a garden. We go through times when we cannot seem to get anything to germinate or sustain what we already have. Pests and fungus multiply in silence beneath the soil today, as they did in Haggai's day.

"I struck all the work of my hands with blight, mildew and hail, yet you did not turn to me," declares the Lord." Haggai 2:17

The dream for a beautiful garden never includes fantasies of hard work. First, there is edging that defines a boundary. Then there is sod removal, digging, and pulling up deep anchoring roots. In and out of seasons, each showpiece needs weeded around, pruned and care. The trash can't be left lying around. So roots, trimmings and weeds are stacked into piles.

This trash can be very useful, and it never runs out. All trimmings are great for compost. Small miscalculation with the weed eater, pruners or mower can be costly. Add them to the pile. On top of those, branches of trees are often bountiful after a storm. If we are lucky that will be the worst of the damage. It is possible our biggest loses come from the smallest predators. Slugs and bugs invade a garden with little mercy.

"I will prevent pests from devouring your crops, and the vines in your fields will not cast their fruit,' says the Lord God Almighty. Then all the nations will call you blessed, for yours will be a delightful land," says the Lord God Almighty.
Malachi 3:11

Even if if there are years we don't struggle with bugs, still seasons come and go. Winter is promised after fall and drought sets into every life at unpredictable points.

"As heat and drought snatch away the melted snow, so the grave snatches away those who have sinned… the worm feasts on them; evil men are no longer remembered but are broken like a tree."
Job 24:19-20

Yet God knows what the world holds for us and is able to orchestrate new blessings. He did not give us a garden and leave us helpless. The earth speaks to His vigilant care in that gardens are equipped to reuse and remake. He had a plan for what would go wrong in the garden and within a man. It will not do for us to show the Lord the garden and not the garbage. We can try to keep the rotten stuff from our neighbors and friends, but He wants it all. He can remake the compost when we hand over what is broken and decaying within our lives as within our hands. In Raising Kids to be Adults, Donald Joy says, "God is our first and best confidant. So surrender, give up, and make a bold appeal for God to speak to our deepest motivations and longings to find secrets which we have hidden from ourselves." Only the Lord can render our mistakes impotent and make them useful.

The same God who walked each evening with Adam hopes each day to walk with us through whatever it is that makes up our gardens. The details of our lives, I am sure, are stark and pathetic compared to the Eden He meant for each of us. But, He is not afraid of our failures. He planned for them. He took them as His own and died to restore what we lost.

"When I consider your heavens, the works of your fingers, the moon and the stars which you have set in place, what is man that you are mindful of him . . .?" Psalms 8:3-5

- Do you remember a moment when you overcame the hurts of your adolescence?

- If you could have only one fresh start or do over what would it be?

- Do you have a compost pile in your garden? What do you do with the compostable matters of life?

Organic Matters..

"What a wretched man I am! Who will rescue me from this body of death?" Romans 7:24

did not purposely mean to hurt a friend, but it didn't mean that a friend was not wounded by my actions. She volunteered to water around our church when it was first planted. It was a lot of watering. I put a signup sheet out to get others to help.

had not asked if she wanted help and as an immigrant she was struggling to find a place n our community. She felt very isolated, but watering gave her a chance to feel needed and gave her a sense of place. She found precious time alone with the Lord through it. Because others took over her work, she felt that I did not think she was competent and took her gift for granted. I had not known her heart, but I had not tried to know it either. I apologized.

"Forgiveness is the fragrance the violet sheds on the heel of the one who crushes it." - Mark Twain

Relationships can get messy even when we meant to support one another. Nature is messy. Leaves, petals, branches are always falling. Suckers and overgrowth pile up as we prune. Foliage collapses as roots enter dormancy. What is true for a garden is true for a man's life.

The decomposing organic matter of a life is untidy. Every man that lives and breathes injures things in his path and wounds things in his care. Our imperfect choices, thought life and behaviors, produce weeds, wounds and neglect, even if very few people see them. The most sincere attempts in the morning continually lay down in the evening with some regret attached. It is comprised of everything from our best intentions to the consequences of our arrogant, vengeful nature. Made up of what we cover or suffocate, sin is seductive. It isn't that one decision wrecks our lives, but small cumulative decisions piled can cause us to stumble, become misguided and burdened.

Gardens begin after what existed before them have been torn out, broken apart and striped away. The debris does not lessen in quantity with time. There will continue to be work overflowing from new life as it takes root there. In The Four Loves C.S. Lewis is the first person who put this into words for me. In his own way he says...
. A garden only comes into being because someone has taken time to plant it... A garden only remains a garden, because someone has taken time to maintain it. He tied the formation of a garden to the establishment of a relationship and the care that follows. What we do with the throwaways is as important as the care we take to establish a garden.

God has a plan for life's throwaways. As foolish as the idea may seem, the very things that assault and overrun a garden can bless it. Made up of rotten greens, leaves, worm droppings and other unmentionable waste products, the best thing a gardener has to give back to the garden is composted garden garbage. For all the romance a garden offers its visitors, for all the sweet life-giving fruit that it produces, compost is the gift a gardener returns. A fertilizing mixture of decomposing organic matter. It is made up of both life and decay. It is useful only after it has rotted.

"And we know that in all things God works for the good of those who love him, who have been called according to his purpose." Romans 8:28

With every area of our lives surrendered to God, this verse assures us that God takes even failures, wounds and regrets. Then He brings restoration. Compost is not a gift you'd give for Mother's Day, but it is a gift to a garden. A garden asks for nothing better.

"But God chose the foolish things of the world to shame the wise; God chose the weak things of the world to shame the strong. He chose the lowly things of the world and the despised things – and the things that are not – to nullify the things that are, so that no one may boast before Him." 1 Corinthians 1:27

Growth is an important indicator of life. As we grow, we celebrate milestones marking that change. We are assured of areas of defeat as we develop. Our lives take corrupt and wild paths. Unmentioned as they may be, rotten, miserable, dreadful transgressions accumulate daily in life's garden. Romans assures us this is true for even the best of men.

"I know that nothing good lives in me, that is, in my shameful nature. For I have the desire to do what is good, but I cannot carry it out." Romans 7:18

Each day produces leaves of memories that flourish in green, and then glow in vibrant color as they become yesterday. The moments slowly fall away and crinkle up as they get tucked into our memories. Life branches in one direction and then another, never to return to what it once was. Branches die and life sends out suckers that steal our time and energy. Even the choice stock the Lord plants can turn corrupt.

"I had planted you like a choice vine of sound and reliable stock. How then did you turn against me into a corrupt, wild vine?" Jeremiah 2:21

The compost of life is made up of things we need to let go. Life comes with mistakes, failures and heartbreak. Some of the wounds are accidental. Others are self-inflicted. At some point we're all rocked to the core. Whether we run or crawl from what changes us forever, letting go of the thing itself is sometimes easier than letting go of what it did to us. Compost forms when we look back and laugh at the misfortune. It loses its hold. Then the hugs and warm embraces of relationships that come after forgiveness and restoration fertilize the soul.

The Creator is a God of second chances. He uproots, tears away and weeds. He prunes and transplants. He heals and restores us.

"This is what the Lord says: Your wound is incurable, your injury beyond healing. There is no one to plead your cause, no remedy for your sore, no healing for you . . . But I will restore you to health and heal your wounds." Jeremiah 30:12-17

• Has a misunderstanding once dealt with, ever strengthened a friendship? Been a reason to laugh, bonded you closer?

• Good memories like flower petals, failed expectations like broken branches, ever changing schedules and another's hurt feelings (and our own) like weeds pile up in every relationship, how often do you take time to tend to them and pull them away?

Trimmed and Pruned...

Is it just me or does it seem like there are less and less climbing trees around today? We planted a Crimson King Maple tree at our farm. When it was young, mom convinced dad to leave the lower branches. It was always hard for dad to mow around, and I'm pretty sure, it would have saved him trouble to cut branches back. It was too small when we were kids; but before the farm was sold, my son was able to climb from the bottom branches high into the tree. It was an amazing adventure for a little guy, and the tree was beautiful! The tree grew in the right place and patience allowed it to be a blessing in its time. If the lower branches had been trimmed, something beautiful would have been diminished.

Most trees grow taller and broader than other plants, and in nurseries. They come labeled with their expected size. If used correctly, this information empowers a gardener to plant them in a place where they can branch out and become exactly what God intended them to be.

How and when growing things are pruned in a garden is important. It is often necessary for healthy growth. The same is true for our lives. When looking for a place to flourish it would be nice if people came with labels. It is hard to predict what God-given nature or shape God will bring out in a person. But often it is easy to tell when they are growing in the wrong place. They look haggard and beaten down. They appear as over trimmed as some of the butchered plants surviving in front of people's homes. If a person grows in a place God has cleared for them they never need to be cut down to size.

Pruning is needed for many reasons. It can shape and encourage growth while getting rid of anything undesirable. It can change a plant's shape and sometimes change a plant's nature. Nipping the buds at the tip of a stem or root forces dormant buds back on the branches to grow and helps a plant to fill out.

I never think of roots when I think of pruning. But Bonsai trees seem magical, because they are cared for in this way. Attentive gardeners dwarf them by pruning roots and pinching their branches. Containers keep them small. Its caretaker sculpts it to a desired shape, because they take time to understand their nature. It is the only way they learn to know what to expect from growth or how to prune correctly.

This works for all trees. For example, apple trees produce more fruit on fewer branches if trimmed to open its top to light. Most roses need to be pruned constantly, but Knock-Out Roses flower more if the rose hips are not cut off. But understanding is plant specific. There is no one size fits all approach.

God prunes us to shape our growth much like that. He cuts away overgrowth. Shaping our lives, He keeps us from becoming unruly when we allow Him. What God prunes in our hearts keeps us from pride and arrogance. He trains growth in the landscape of His love. He asks us to cut away suckers that sap strength and trims our roots to train our growth. He always makes our lives fuller and His people more fruitful when He does.

"He cuts off every branch in me that bears no fruit, while every branch that does bear fruit He prunes so that it will be even more fruitful." John 15:2

Pruning is rarely easy. Working around lower branches covered in dead and broken sticks is frustrating. I hate when a stick swings around and snaps me in the face, my hair gets caught in a dried twig, and despise it when I poke my eye with them. Sometimes it is hard to reach the branches. At other times it is hard to cut away attractive vegetation. Then you have to wait for new growth even if we know it will keep the plant healthier. There are a lot of different ways to prune. In our electric and gas powered age we have hedge pruners and chainsaws. A person's fingers work perfectly pruning boxwood, though it takes a long time. Hand pruners, scissors, saws and loppers are great tools. Always aim to make as small a wound on the plant as is possible with no tearing.

We all have disposable things that interfere with how God would shape our gifts. When He changes the direction of growth, what lies dormant within our branches branches out.

"Go through her vineyards and ravage them, but do not destroy them completely. Strip off her branches. . ." Jeremiah 5:10

Pruning cuts away what drowns out our Savior's voice. God uses it to shape us to His will. He becomes the desire of a heart when everything superfluous is cut away. Even at times, blooms need to be cut and put in vases. Then they should be shared. That is why God gives us gifts.

Some good things can grow to be ugly or even hazardous because their time has passed. A shrub that is never trimmed can outgrown its beauty and usefulness. They also begin to soak up all the water and block light from growing things around them. In life prideful gifts, manipulation and selfish choices need confronted too. A strong rope, backhoe and woodchipper too would be perfect tools against them. It's not just mistakes we recognize easily that may need this.

Accomplishments potentially ignite an overgrowth of arrogance. We soak up all the water and light, and forget what lies beneath our branches.

Pride and arrogance take many disguises. They are fueled by the beauty and splendor we love about ourselves.

"In his pride the wicked does not seek Him; in all his thoughts there is no room for God." Psalms 10:4

There are many things that can grow in the landscape of our lives that are best left undisturbed, like the Crimson King Maple I remember. However, gardens are small contained spaces and many trees and shrubs benefit from the attentions of a gardener. The same is true for many elements of life. Details matter. Understanding the growth habits and gifts of growing things makes a world of difference. Evil cannot be allowed to establish itself where it chooses. It is not finicky; it feasts on the best and worst things about us. Mistakes and successes form us.

"God has made us for himself, our hearts are restless until they find rest in Him." St. Augustine

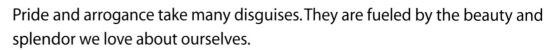

• Do you have any memories of a tree that branched out and you were glad was never pruned.

• Can you think of any part of your life that has needed pruned, but not completely removed?

• Have you ever cut a shrub back and watched it grow back more beautiful than ever?

Compost Pile...

"For I know my transgressions, and my sin is always before me. Against You, You only, have I sinned and done what is evil in your sight." Psalms 51:3-4

A compost pile doesn't look pretty. Its smell is no more appealing when kitchen scraps collected for a week are thrown on top even if they are not yet slimy. But it is made up of all that has been pulled away, cut down and thrown out.

The compost pile of life is no different. It is made up of things for which we have asked forgiveness. It doesn't make our lives prettier or add an aromatic aroma, but it can make it better.

A gardener cares for a compost pile by turning it over and over. This helps it rot more quickly and completely. Turning it exposes every segment to sunlight. When it doesn't rain it may need watered. The water helps the center of it to heat up and breaks down what makes up its content. Time then is needed for it to rot fully. The four seasons here in Pennsylvania bring summer heat which cooks it. The freezing of the winter which breaks it apart is helpful. Turning it again helps to disintegrate it further until it turns into soft, workable soil with no foul odor and fertilizing properties that are unmatched.

"If we confess our sins, He is faithful and just and will forgive us our sins and purify us from all unrighteousness." 1 John 1:9

To me turning compost is much like the action of bringing our sin and mistakes before God. It is acknowledging failures and being honest about weaknesses. It is releasing them back to the rot and decay from which they came. It recognizes their character and puts them in their place.

Living Water and the Light of the World are catalysts for its renewal. They expose what lies in the darkness as in 1 Corinthians.

". . .wait till the Lord comes. He will bring to light what is hidden in the darkness and will expose the motives of men's hearts. . ."
1 Corinthians 4:5

We misunderstand what God uses for restoration when we offer him only joys and triumphs. The best image management in the world cannot expunge the truth. God is not deceived by disguises. Asking for healing for our bodies and failing to name sin before God produces what neglect encourages us. He wants to take the mistakes and throwaways of life and transform them. He alone can do that well. He does not want us to cherish anything good in our lives more than Him nor harbor any evil. He wants the fallen crinkly leaves and frost bitten foliage. God wants weeds, not to preserve or celebrate them, but to smash them to dust and remake them.

"I beat them as fine as the dust of the Earth; I pounded and trampled them like mud in the streets." 2 Samuel 22:43

ime has a way of changing what we remember. Turning the sin pile of our lives may bring up memories we've tried to forget. Each turn f the pitchfork, brings to light oversights. Each confrontation thwarts the possibility of repeating past mistakes. It keeps our memories onest and invites the Lord to show us what is overlooked. It breaks the past's grip on our hearts.

here will be things on the compost pile that we don't mind leaving there. Some things are harder to let go of, but once we see them ring in their own rot, it gets easier. Some things we only accept as trash when they are smashed to pieces and covered in a fuzzy decay.

he broken down and disintegrated remnants when they have stopped rotting are a gardener's gold. In the same way that weeds are iled up and allowed to rot, the Psalms tells us that God wants the sin of our lives forfeit.

"The sacrifices of God are a broken spirit; a broken and contrite heart, O God, you will not despise." Psalms 51:17

Then God amends what is broken. His power is made perfect when weaknesses are confronted.

". . . My grace is sufficient for you, for my power is made perfect in weakness." Therefore I will boast all the more gladly about my weaknesses, so that Christ's power may rest on me. That is why, for Christ's sake, I delight in weakness, in insults, in hardships, in persecutions, in difficulties. For when I am weak, then I am strong." 2 Corinthians 12:9-10

f course, in a garden as in life, we need not compost at all. We can hide what is gly and ignore it completely. But pretending that nothing is wrong only ensures at it will rise to haunt us. Garbage is made up of things we hide. Restoration can't e found in the poses of perfect family pictures or caked on makeup. And I admit I se them, but Snapchat filters are controversial for anyone authentic. Empty giggles cho through emptiness and there is no filter for that. Ecclesiastes tells us laughter des nothing and is cruel to a world that longs to know truth.

"Like the crackling of thorns under the pot, so is the laughter of fools. . ." Ecclesiastes 7:6

Sin loses the power to harm us and gains the power to heal when it is exposed before the Lord. As we lay our weaknesses at his feet, He turns them over in our hearts and changes how we see them. Sin, when it is dealt with, can heal the deepest parts of us and restore even the bones we have broken skirmishing with the world.

"Let me hear joy and gladness, let the bones you have crushed rejoice. . . Create in me a pure heart, O God, and renew a steadfast spirit within me." Psalms 51 : 8, 10

Hardships or harsh consequences become a part of our soil. Like relationships we wrestle through that become more precious because of the struggles we went through. The compost pile is a collection of thing we have let go of, but yet if confronted, we become stronger trhough what we learn. The wrestling, tearing away, and all out battle then becomes the very thing that strengthens and enriches us.

• Are there things in your life that you struggle to let go of?

• Have you exposed things and been thankful when God released you from their hold?

Repentance and Forgiveness...

"Bear with each other and forgive whatever grievances you may have against one another. Forgive as the Lord forgave you." Colossians 3:13

In Romans 2:4b I found this statement. Do you not realize …"…God's kindness leads you toward repentance?" There is always hurt and loss in this life. In those moments, kindness brings a desire for atonement. Gardening in those moments promises restoration.

I looked up repentance in the Bible dictionary. The Greek word "Metanoeo" used in the New Testament refers to the changing of one's mind and purpose as a result of after knowledge. Psalms 51 shows us true repentance in this way. "The true penitent is conscious of guilt (Ps. 51:4, 9), of pollution (51:5, 7, 10), and of helplessness (51:11; 109:21, 22). Thus he apprehends himself to be just what God has always seen him to be and declares him to be. But repentance comprehends not only such a sense of sin, but also an apprehension of mercy, without which there can be no true repentance (Ps. 51:1; 130:4)."

I tell my kids all the time, "You are going to make mistakes. The trick is to recognize when they have made them and to learn from them." I tell them why they need to be honest about them; it is important to apologize and to make things right if they can.

True repentance is a continual change of behavior. In its simplest, sense it is an attempt not to cause another injury. It is not always easy to do.

Forgiveness stands united with repentance. Forgiveness is a willingness to exonerate. Repentance then drags away the wreckage of life. It composts hurts we have received and those we have inflicted. Because of it the Lord can replant. If anything can be grown from the ruins, all these things need to happen.

"On the day that I cleanse you from your sins, I will resettle your towns, and the ruins will be rebuilt. The desolate land will be cultivated instead of laying desolate in the sight of all who pass through it. They will say, "This land that was laid waste has become like the Garden of Eden; the cities that were lying in ruins, desolate and destroyed, are now fortified and inhabited. Then the nations around you that remain will know that I the Lord rebuilt what was destroyed and have replanted what was desolate. I the Lord have spoken, and I will do it." Ezekiel 36:34-36*

God's forgiveness is the only hope we all have. He remembers not our sin.

"As far as the east is from the west, so far has removed our transgressions from us. . . for He knows how we are formed; He remembers we are dust."
Psalms 103:12, 14

Do not to let the shortcomings, wounds and neglect of life pile up. Our Creator uses forgiveness in much the same way as a gardener uses compost. He uses it to restore joy, song and dance.

"You turned my wailing into dancing; you removed my sackcloth and clothed me with joy, that my heart may sing to you and not be silent. Lord my God, I will give you thanks forever." Psalms 10:11-12

Too often productive energies are exhausted hiding trash. When we deal with it through forgiveness and repentance, uglier things like insecurity, anger, and low self-esteem can be avoided. Ugly things are the only things that grow from sin. The smell from them lingers. Hoping no one notices the stench doesn't change the fact that it keeps us up at night.

Compost definitely does not belong in the front yard for neighbors to muse over and complain about. Wearing garbage on our sleeves does little to promote healing. It is ok to protect our secrets and keep them from those who would do anything other than hold us accountable. Compost piles are hidden because they are ugly. An isolated spot is a good place to work with one. It keeps the smell away and the pests it attracts at a distance.

Only our closest companions are invited to stay awhile in the backyard of our lives. Those who can be trusted to turn the compost with us will turn it to help cover our shame and destroy its infective power. Those who love us want to wipe away what would stir up fresh sorrow within us. And they will admit freely that they have their own compost pile in their backyard.

"Above all, love each other deeply, because love covers a multitude of sins." 1 Peter 4:8

• Has God's kindness led anyone you know to repentance?

• Has the stench of hidden trash in your life ever kept you up at night?

• Do you know people who keep their "compost pile" in their front yard?

Fertilizer...

My husband interacts with a lot of students. One of the things he says about students is that the outstanding kids all have one thing in common. By outstanding, I do not mean the smartest or the most popular. I mean the ones he is confident will be successful and looks forward to knowing as adults. The quality that separates them from the others is that their parents spend quality time with them and vigilantly watch over them. These kids make mistakes, don't misunderstand that. The difference is that their parent points them out. They are nurtured, trained, and discipled. I love the book 12 Rules for Life by Jordan Peterson. Rule Five in the book is, "Do not let your children do anything that makes you dislike them." I love this advice. That they learn how to improve from their mistakes marks a difference in them.

As a garden is fertilized by compost, experience shapes

humanities story and is our teacher. Fertilizers work because they add something that is deficient. Growth is lush and healthier, blooms and fruit are tastier and larger, because of what is returned to the foundations of what is grown.

I was a better mom, and my kids were cared for by a happier woman, because my mom and other moms taught and encouraged me. I was helped the most by their stories of failed attempts. New experiences in life are sweeter and far less scary when a mentor shares their life and its compost. Everything from the teen years, dating, marriage, horseback riding, gardening or working in any job is easier with good counsel. The best counsel includes the details of good and bad experiences. In God's hands, mistakes no longer causes harm, but when broken down can **nourish and amend the soil of life. It is reborn.**

Compost, when fully rotted, is a natural fertilizer. It feeds and encourages healthy growth. It adds both basic building blocks of life and flaky organic fibers that help keep the soil loose for root growth in clay soil or holds moisture in sandy soil. It blesses the garden and the one who gardens. Our life's experiences, when fully left at the Lord's feet, add building blocks to our lives and to others.

Fertilizer makes the soil capable of producing abundant crops. It causes it to be productive. We fertilize if we want that which we have worked hard to be healthy and fruitful. God uses the composted sin of life in the same way. He makes our lives fruitful, encourages beauty in us, restores joy in our sorrow, and ignites praise in the place of our despair.

"To all who mourn in Israel, He will give beauty for ashes, a joyous blessing instead of mourning, and festive praise instead of despair. In their righteousness, they will be great oaks that the Lord has planted for His own glory." Isaiah 61:3

There is one type of fertilizer that cannot be manmade. It is divinely made by the heavens. It is lightning. When lightning strikes the ground it adds nitrogen to the soil. I was reminded of this when I read Psalms 135.

"I know that the Lord is great, that our Lord is greater than all gods." "The Lord does whatever pleases Him, in the heavens and on earth, in the seas and all their depths. He makes clouds rise from the ends of the earth; he sends lightning with the rain and brings out the wind from His storehouses." Psalms 135:5, 7

I think God gave lightening alone that power, because some things we have to depend up from His hand alone. Lightning can be terrifying. It cannot be controlled. Our dependence on God can be just as frightening if we are honest. It is best to stop moving, crouch in a submissive position and pray through the storm.

Any given place is exposed to all the weather the seasons can muster. But you can never be sure day to day what gifts may be discovered there. The sun always sets even if you can't always see it. Somethings are worth waiting for. But you have to be there to experience the gift. People are like that. We go through seasons. Believe in the sun behind the clouds and you have a better chance of being there when it counts. Sweet friend, sin piles up in everyone's life. For all the cherished dreams of our youth; for all the best intentions we

had, our accomplishments come with shortcoming, injury or sorrow. Life is messy. No life is pretty around every corner and in every season. We cannot ignore trouble and expect it to go away. Sin goes to seed and rots where it falls. Sorrow stunts growth. Wounds fester and diseases spread. Take a rake over them, a shovel, pruner, a chain saw and some common sense. Cut it off and let it go! Take the garbage out! Compost what you can.

Compost's fertilizing power is useful only when it is worked back into the soil. In the same way, the lessons we learn from our failures are only beneficial when we remember them and when invited to do so share them. The decomposition of sin helps us to let go of regret and gives us something useful to do with it. When it is through rotting in the Lord's care, it has fertilizing qualities that encourage new growth in us.

"For the foolishness of God is wiser than human wisdom, and the weakness of God is stronger than human strength." 1 Corinthians 1:25

Not every part of our lives needs to be an opened up to others, but in some places, and at appropriate times, consider sharing your compost. Sharing your life with others can lay fertilizing mulch over life. Vulnerability breeds vulnerability. If we are honest, it will encourage others honesty. The only way for another to feel safe with us is if we are safe. We only have the right to share our own "dirt."

• Have the stories of others in your profession, field of interest or parenting helped you through life?

• Has a lightning strike ever invigorated the soil of your life?

• Have you shared parts of your life hoping to enrich another person's?

Second Chances....

"Love is never wasted, its value does not rest on reciprocity." - C.S. Lewis

Vacations are like second chances. They compost daily life. We walk away from all our hard work. Escape from reality and refocus. When we stand beside the ocean we are reminded how small we feel. The sunset then reminds us of a God bigger than the sun, deeper than the ocean and endless as the sand.

Our God is the God of second chances. God lifts us from the garbage dump.

"He lifts the poor from the dust and the needy from the garbage dump. He sets them among princes, placing them in seats of honor. For all the earth is the Lord's and He has set the world in order." 1 Samuel 2:8

Cultivating any garden requires more of us than we may have first think is necessary. At times it requires more than we believe possible. When we are rooted and established in the right place, our lives can be filled with fruit that surpasses understanding.

"... And I pray that you, being rooted and established in love, may have power, together with all the saints, to grasp how wide and long and high and deep is the love of Christ, and to know this love that surpasses knowledge - that you may be filled to the measure of all the fullness of God." Ephesians 3:17-19

Seeing compost as a precious commodity may be a matter of perspective. Or maybe… it is God's sense of humor. He keeps us grounded to the truth of our lives. Whatever He meant us to understand from it, it works! What it can teach reaches as far as each individual who would try to understand it. When His love consumes us, He restores us like fire restores the fertility of the land after a brush fire.

". . . our 'God is a consuming fire." Deuteronomy 4 :24

God works with living things and decaying things alike. While it will not, hurt the garden if compost is not added to it. It will never be blessed by it either. God is about making our garden better. He is about making us better. Composting the sin of our lives doesn't change Him. Composting does change the world around us! It changes us! It gives us a remedy for a broken world.

"A cheerful heart is good medicine, but a crushed spirit dries up the bones."
Proverbs 17 :22

After the debris has rotted in a garden it is reborn. It has a fresh earthy smell. A gardener can hold in his hand the beauty God makes from ashes. A man can hold in his heart love, joy, peace, and hope God gives us despite our failures. It makes new experiences richer and more abundant. It helps us laugh at ourselves. It makes our stories useful in leading another away from the same mistakes. Its new qualities can then fertilize dreams. The earth can sing again.

"Sing to the Lord a new song; sing to the Lord, all the earth." Psalms 96 :1

A Garden Observed: CULTIVATING A LIFE

The most interesting person in any room is the one whose life has been filled with experiences and who has learned from and built on both the good and bad. It's the person who can laugh at themselves and help others to compost their mistakes in the same way. The least interesting person in the room is the one who has taken those trials and heaped them on their own back or hidden them in silence and disappointment. Because when we fail to confront what is wrong. We start a war with ourselves. "When you have something to say silence is a lie," as Jordan Peterson would say. Neither they nor those around them will ever grow or be nourished from silence. A life's landscape is formed by the storms, upheavals and seasons. Did we give them to the Lord? Learn from them? Change from them?

Compost is what we gain from looking back on life, especially the hardest parts, and being able to point to the beautiful. It's seeing a neglected farm house, recognizing a struggling family and yet rooting to every precious memory that made survival worthwhile. It's not pretending that life wasn't hard. It's the ability to hold on to what is lovely, good, precious and valuable even after it's been lost and changed. I hope that's what this journal helps others to do.

Let us never cease to be amazed by what God can use to make something beautiful. He is not afraid of what makes us who we are. But He loves us too much to let us remain in our own rot and decay. He can make from it more than we dare to imagine.

"Now to Him who is able to do immeasurably more than all we ask or imagine, according to His power at work within us, to Him be glory in the church and in Christ Jesus throughout all generations, forever and ever! Amen." Ephesians 3:20-21

f God uses the decomposing organic matter of the earth, there is hope He can restore our lives as well. Then and only then will we be eady to plant the seeds of new beginnings.

"But whatever things were gain to me, those things I have counted as loss for the sake of Christ. More than that, I count all things to be loss in view of the surpassing value of knowing Christ Jesus my Lord, for whom I have suffered the loss of all things, and count them but rubbish so that I may gain Christ, and may be found in Him, . . . that I may know Him and the power of His resurrection . . ." Philippians 3:7-10

• Do your think God has a sense of humor?
• Do you think He wants you to have one?
• Is there anything about you, you know God wants to change?

GARDEN
Always

Concerning Stones...

"There is a time to scatter stones and a time to gather them." Ecclesiastes 3:5

Like many farmers' kids from Somerset County, I remember gathering stones from freshly plowed fields and unloading them along tree lines. They were added to piles I am sure had been started generations before. It was muddy, hard work. Every time rain fell, rocks grew like crops across the soil's surface. As the soil dried in the sun, a never-ending sea of stone surfaced and waited to be picked up. At times suspicious of a cruel joke, I imagined my parents scattered the stones while we were away, plotting to cut into our play time.

As a grown up, I now know there was no cruel joke about it. Life is filled with obstacles that must be circumvented, removed, and surmounted. They can be burdensome like rocks. However, it is also possible for them to point us in new directions. With a fresh perspective mounds of stones can turn into fields of treasure.

The most memorable stone work for me is that which I've had the privilege to observe being assembled. Watching a stone mason assemble walls, sculpt waterfalls and define ponds can make you believe anything is possible. Magic craftsmanship sets ablaze a love for all things stone within me. It could not be quenched by all the stone piles of Somerset County. A treasure trove of architectural elements lies waiting in those stone pile, some of which may be more than 100 years old.

Deciding what to build and shaping a collection of stones into a landscape is one of my favorite parts of gardening. My heart races when a day is available to gather them. My husband and I drive along farm fields, and with permission from each farmer, we stack stones on our truck. My husband grumbles a bit when we pick up a heavy one together, but as I pick each one up I feel hearts dancing within them. Something deep inside of me believes they have been waiting to be useful since the day they were formed.

I never get enough on my truck when the bed starts to sag too near the tires, and we have to stop. Returning home, I always have to narrow my ambitions. It is a bit daunting to think their weight, our strength and my imagination are the only limits to their potential.

As simple as a single stone can seem, there is so much that can be done with them. They have an ancient and lasting permanence and come in all shapes and sizes. In a garden, river rocks can form dry river beds that don't have to pretend to flow with enchantment. Their unique coloring, smooth texture and varying sizes add interest around colorful blooms and lush green backdrops. As the stones age in the landscape, their patina becomes more and more attractive. Man made elegant statues of sculpted stone speak volumes with their silent lips. Boulders avail stoic beauty. They can be used as benches and backdrops. Rock walls retain boundaries for forest and flower. Those walls can keep water away from a home or keep pets and children safe. Stone paths direct visitors around a rock border harboring a campfire ignited by songs and stories.

I remember my dad's frustration as stones broke the plow and hindered his work in the fields. I never enjoyed all that work under the sun either. I never planned to pick up another stone when I got away. But in truth, most of life is hard work that always seems to cut into playtime. No matter where you live work struggles intermixed with risky ventures. But the hardest things in life can often be the most precious. It's important to enjoy the journey or the rocks start to weigh a person down. I thought our family farm and that place in time would always be there: fields to wander through for hours, haylofts for hiding places and baby animals to kiss. The barn windows glittering at sunset, and treasure was as simple as a story and dinner together around a table. It doesn't last.

Rather than jewelry, I love when my husband brings me a heart shaped rock home from a fishing trip. They are not always easy to find even if they are free. They tell me he has been thinking about me when he does. My son has learned from his example. He once carried a heart shaped river rock he found

two days in a canoe and over his shoulder home for me. I like to think it is because I am still his girl, though I won't be for long.

I first saw a collection of stone hearts at a cottage in the woods. A ledge under a back porch was lined with rocks stacked and leaning. They were big and small and shaped like hearts. I don't know for what purpose the cottage owner collected them. But they reminded me of a meeting with my God, and the choice I make each day about the heart I open or close to Him. They remind me that my life draws others to the Lord or blocks Him from their view. Since that time I have collected stones in the shape of hearts. Some I pull when working in my garden, most I collect on walks or when I am gathering stones for a project.

There are rocks I do not like. I do not rejoice over little stones washed up on the surface of a freshly plowed vegetable garden. Nor do I love the rocks buried beneath the soil that a shovel hits and grinds against. Digging for them and pulling them up with great effort does nothing for my enjoyment.

I had to ask the Lord, "What there was to learn from all the little rocks?" If God has a purpose for all things, what do the stones, washed clean in the rain and dried out covering the ground, have to teach us. Stones collected in buckets and thrown into a tree line or ditch seem pretty useless. I honestly wondered, how are they different than the weeds, we use for compost? Compost can be reused in a garden. Little stones are never planted or returned to a garden. I found this verse…

"Like rocks brought up by the plow, the bones of the wicked will lie scattered without burial." Psalms 141:7

I know it may seem like I am digging for meaning and the truth is, I am.

Rock-strewn ground is as abundant in life's garden as frustration. Small stones on a garden's surface add no value to it. If they are left lying, every year when the soil is turned they resurface. They injure joints, stub toes, trip us, stifle seedling growth and break shovels. I believe God showed me they are akin to idols. When they are removed, they need thrown out. When collected and misused they remind me of the idol the Israelites made into a bull. It was made from a mineral of the earth that was misused.

"They exchanged their Glory for an image of a bull which eats grass. They forgot the God who saved them who had done great things…, miracles, awesome deeds," Psalms 106:20

Stones are heavy and have no power of their own. When they become idols, they lead us to terrible ends.

"Those who consecrate and purify themselves in a sacred garden with idols in the center .. . will come to a terrible end," says the Lord. Isaiah 66:17

Sometimes it's the wrong person we have trusted. Sometimes it is the wrong god we have believed. Even if made out of beautiful things, it does not make it a treasure.

"Cursed is the man who trusts in man, who depends on flesh for strength and whose heart turns away from the Lord."
Jeremiah 17:5

The Lord has purposes in both great and small matters of our lives. We should be looking for the value He gives when we build anything.

Now give careful thought to this from this day on – consider how things were before one stone is laid on another in the Lord's temple . . ." Haggai 2:15

An idol is as an image of something. It is a material substance that is worshiped, with false admiration and devotion. Idols are fakes and phonies. When they draw our attention and adoration away from the Lord, "…They can do no good, nor can they rescue you, because they are useless." 1 Samuel 12:21b. Whether small enough to fit in our wallets or as large as boulders, they have the power to crush us. We attribute value to them they should never be given. If we gather them as treasure and plant them, when we are in need of harvest, there will be none.

In Joshua 7, there is a man named Achan who hid idols under the cot in his tent. I don't think that his actions were all that different from what we hide in our lives to this day. Such secrets things are not written down, but buried, chained, and caged in our thought lives. Often they are things that we would prefer to keep underground. They are secrets that offer no power, beauty or growth.

Achan was put to death for hiding his bronze idols. His family was stoned with him. Men lost their lives in battle. Families lost their fathers, brothers, friends, and sons.

"... At this the hearts of the people melted and became like water." Joshua 7:5

The idols could not talk, walk or move, and had no power other than what Achan gave them. Idols are helpless stones that offer us nothing, like the smallest secrets we keep from God. They separate us from His blessings. Like stepping stones, they guide our hearts away from the Lord. This is why God doesn't want anything hidden from Him in our lives. He would that we would expose, uncover, and come naked before Him.

"This is what the Lord says ... "Break up your unplowed ground, and do no sow among thorns. Circumcise yourselves to the Lord, circumcise your hearts."
Jeremiah 4:3-4

• Have you ever gathered something someone else threw away?

• Is there any craft that excites your soul? How does it affect your foundation?

• Has God ever revealed an idol in your life?

Stone and Rock...

I wanted to know so I looked up the definition of stone in the dictionary. A stone is defined as a hard mineral substance, small rock or gem. I looked up rock as well. It was simply defined as a mass of stone. From such a description, a rock or stone may not seem to be anything exceptional or even useful. But, how we define the stones we treasure matters greatly.

"... Look to the rock from which you were cut and to the quarry from which you were hewn." Isaiah 51:1b

Stones have unforeseen and unpredictable strength. The pebble in the runner's shoe is one example. The little stone is a nuisance. It hinders the journey and was more troublesome than the boulder or mountain in every step he took. The power of the stone that David used to defeat the giant in 1 Samuel 17 may be more to the point or even the millstone that killed Abimelech in Judges 9:53. In Deuteronomy 8:15 and in the Psalms 105:41-42, God provided water for his people from rock. The Bible says that if we do not praise the Lord, the rocks will praise Him in our stead.

"'I tell you,' He replied, 'if they keep quiet, the stones will cry out.'" Luke 19:40

But there are predictable things we expect from stones. Stones can be used to build walls. If those walls are beneficial or not may depend on the reason they are built. The Lumineers have a song called Walls, the lyrics are, "all around your island there's a barricade, that keeps out the danger, that holds in the pain." Walls can do just that if built for the wrong reasons.

I'd love to see the stone fence rows in Ireland. We edge our gardens with thick, sturdy stones. Many gardeners use them to delineate paths, and I love the texture and color pea gravel adds as it fills in between the cracks. Enchantment in stone is a stone archway. I dream of a Moon Gate like the one in Dare to be Wild, the movie about Mary Reynolds. I'd love to have such an arch just to walk back and forth beneath it. Big boulders or landscape stone add texture and new color to a garden bed. If lucky, one big enough can be used for a bench or even serve as a place on which to lie and warm yourself in the sun. The rolled stones have their own magic as sunlight reflects off a stream flowing over them. I have two dried river beds flowing from my brick home. One cascades down a slope and fills the top of a bed held by boulders. Except for the lack of water and sound, it looks real. The huge boulders in the middle of a raging river captures respect. Nothing but stone can withstand such power. Manmade stone, like cement or concrete, are important. Statues in a garden and pools are all made of this useful element. Their presence speaks volumes in their silence.

"Therefore everyone who hears these words of mine and puts them into practice is like a wise man who builds his house on the rock. The rain came down, the stream rose, and the winds blew and beat against that house; yet it did not fall, because it had its foundation on the rock. But everyone who hears these words of mine and does not put them into practice is like a foolish man who built his house on the sand. The rain came down, the stream rose, and the winds blew and beat against the house, and it fell with a great crash." Matthew 7:24-27

The Bible talks about a bunch of different types of stones. There are altars of stone, millstones and throwing stones. The Ten Commandments were written on stone tablets. Elijah hid in stone cleft of rock as God's glory passed. God is a rock, our ROCK. The Bible says it is Jesus is a cornerstone and stumbling block. Peter was the Rock, on which Jesus builds His house.

Found in these verses: (Hosea 10:1-2) (Mark 9:42) (Lam 3:53-54, Acts 7:57-58a). (Exodus 33:22) (Psalms 71:3) (1 Kings 5:17) (1 Corinthians 3:10) (Matthew 16:18)

• Have you ever hiked in the woods and came upon boulders?

• What impact did they have on your behavior? experience?

• What stone structures do you love?

Reminder Stones...

Stones can be used to preserve and keep records. They may not record years as a tree numbers them in rings, but what is etched and built in stone is not easily erased or dismantled.

"Judah's sin is engraved with an iron tool, inscribed with a flint point, on the tablets of their heart . . ." Jeremiah 17:1

Habakkuk and Joshua called them record keepers, to remind God's people that we will be held accountable.

". . . (Joshua) said to all the people, "This stone will be a witness against us. It has heard all the words the Lord has said to us. It will be a witness against you if you are untrue to your God." Joshua 24:27

"Stones of the wall will cry out, and the beams of the woodwork will echo it." Habakkuk 2:11

One type of record keeping stone was the ancient boundary stones. They were piled in a place to serve as a reminder of God's goodness, mercy, and deliverance. (Joshua 4:2, 5-7) They denote a moment, a place and a meeting with God. (Joshua 4:20-24) The people retold stories and sung about them. I imagine some visits were intentional and some accidental.

Ancient stones laid by our fathers mark a moment we are not to forget, places that remind us of boundaries, failures, and gifts teach lessons.

"Jacob set up a stone pillar at the place where God had talked with him and he poured out a drink offering on it..." (Genesis 28:18) Genesis 35:14-15

Boundary stones designate what is good and important. As we cannot change the laws of gravity, ancient truths are placed to stand sentry. They may appear as simple, as a pile of rocks, but our fathers gathered them for us to cherish. God sculpted them to stand the test of time. They are more precious than gems. Treasure can be found in fields, marked with such stones.

"Do not move an ancient boundary stone set up by your forefathers." Proverbs 22:28

"... (such) leaders are like those who move boundary stones. I will pour my wrath on them like a flood of water..." Hosea 5:10

My favorite stones in the Bible are, of course, hearts of stones ...We are getting to those. I know you are wondering if stones are really that important in a garden. Yes and no. It depends on what you are looking for. How man chooses to build with them or allows them to form life is important. Stone used for shelters, walls and paths have purpose. We can use them to remind us of our God...

"Then Samuel took a stone and set it up... He called it Ebenezer, saying, "Thus far has the Lord helped us." 1 Samuel 7:12

or craft them to shut out His light.

"Their minds are full of darkness; they wander far from the life God gives because they have closed their minds and hardened their hearts against Him." Ephesians 4:18

• Do you know of any reminder stones?

• With the present day controversy over boundary stones share you thoughts. Do not throw stones.

Hearts of Stone...

"As water reflects a man's face, so a man's heart reflects the man." Proverbs 27:19

A stone wall may be built for protection, but one in the wrong place causes confusion and blocks progress. Some men are wounded by stone. Some strengthen their defenses with it.

"What strength do I have, that I should still hope? What prospects that I should be patient? Do I have the strength of stone? Is my flesh bronze? Do I have the power to help myself . . ." Job 6:11-13

Yet the world hardens one man and weakens another. Our hearts are not made like the leviathan. "Its (leviathan's) heart is hard as rock, hard as a millstone…" Job 41:24 How some men end up with stone hearts, broken hearts, not a precious few with kind hearts, I do know. The Lord searches our hearts for answers.

"The heart is deceitful above all things and beyond cure. Who can understand it? I the Lord search the heart and examine the mind. . ." Jeremiah 17:9-10

You never can tell what rocks lay beneath the surface of the ground or a man's façade. Ground that looks rich can be very rocky. Hardened lives often hide from understanding..

"They are darkened in their understanding and separated from the life of God because the ignorance that is in them is due to the hardening of their hearts. Having lost all sensitivity, they have given themselves over to sensuality . . . with a continual lust for more."
Ephesians 4:18-19

What we do know is that God allows rocks to creep up in our gardens, boulders to stand in our way, and that He has His reasons for allowing our hearts to harden.

"Why, O Lord, do you make us wander from your ways and harden our hearts so we do not revere you?" Isaiah 63:17

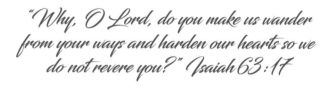

The Bible talks about what God desires to do with our hearts of stone in Ezekiel.

"I will give you a new heart and put a new spirit in you; I will remove from you your heart of stone and give you a heart of flesh."
Ezekiel 36:26

When I read this verse, it was the first time I remember God, catching my gaze and looking me straight in the eyes. He asked me to make a choice about which I wanted. He changed the direction of my life, and the purpose for the stone walls I had been building. He took away the throwing stones I had been gathering and redirected me away from the boulders I had been battling.

My stone heart was something I was content with, though it was not something I rejoiced over. But as Deuteronomy 32:18 shares, I had "…deserted the Rock, who fathered you (me); …forgot the God who gave (me) birth." I never set out to be angry or frustrated. I had set out to break all of Murphy's laws and to defy all the doubts about an insecure, poor farm girl, from Pennsylvania. In one glance, God showed me all of my attempts at happiness had returned nothing. Selfish dreams vanished into cold, hard reality. My heart followed suit."… .waters closed over my head, and I thought I was about to be cut off. I called out to the Lord from the depths of my pit…" Lamentations 3:54

Years later, when the pain I had learned to ignore, would surface it hurt so much more than I remembered. Because once I knew healing, once I knew peace, I never wanted the pain to return.

The stones others threw at me hurt less than the stones I had thrown. They may not have been hard, covered in dirt or fit in my hand. It did not matter. They separated me from God. He allowed my heart to be broken for a reason.

"He pierced my heart with arrows from his quiver. I become the laughingstock of all my people; they mock me in song all day long. He has filled me with bitter herbs and salted me with gall. He has broken my teeth with gravel. He has trampled me in the dust." Lamentations 3:13-14

READ THE REST OF THIS CHAPTER

Our story, like our lives, returns to honesty before the Lord. As the ground turns to stone under the weight of the world so man's heart turns under sin's consequences. Our only chance of restoration is eye contact and a confrontation with our God. The world will disappoint us. We will feel like throwing stones at others to disguise our guilt and point out another's. The Bible warns us in Lamentations 3:65, 4:3, some of us have a veil put over our hearts, a curse on us and become heartless. We dig holes and try to bury the truth about our lives. We pile stones to hide behind. We get stones in our shoes that cause us to stumble. We attempt to climb, boulders that can crush us. They pull us to the ocean's bottom.

"…but as a mountain erodes and crumbles and as a rock is moved from its place as water wears away stone and torrents wash away the soil, so you destroy man's hope." Job 14:18-19

God gave us stones to build protective walls, strong foundations, and places of community. He wants us to cultivate growing things, not stone that turn into idols or hearts that cement us in place. He created a world that flourishes with care.

In Isaiah 9, the people relied on their own strength and not the Lord's.

"All the people will know it... who say with pride and arrogance of the heart, 'The bricks have fallen down, but we will rebuild them with dressed stones; the fig trees have fallen, but we will replace them with cedars. But the LORD has strengthened (their) foes against them. . ." Isaiah 9:9-11a

They thought if they built with dressed stones and cedars, they could rely on their own strength. Yet without the Lord all was lost.

Juxtapose to Isaiah's story, the people of Israel gathered under Nehemiah to rebuild walls around their fallen city under the mocking observations of others. Mocking jeers saying, "...Do you actually think you can make something of stones from a rubbish heap – and charred ones at that?" Yet Nehemiah records, "so we rebuilt the wall all of it reached half its height, for the people worked with all their heart." Nehemiah 4:2,6

God honored their efforts, and they rebuilt the walls from the rubble of their fallen city. If we are not seeking His strength and building on our own we will fail. Even a house build with the best of stones will not be a refuge without Him. The difference was the idols their hearts were serving. Each time we manipulate or forget God's purposes we do so poorly and with consequences. People hurt and hearts harden.

"If you will return, O Israel, return to me." "If you will put your detestable idols out of my sight and no longer go astray, and if in a truthful, just and righteous way you swear, 'As surely as the LORD lives, then the nations will be blessed by him and in him they will glory." Jeremiah 4:1-A

Submitting to the truth about our lives can be just as hard if not harder than smashing stone; as unthinkable for us as forming it into flesh, or even causing water to flow out of it in a desert place. But they can also serve as beautiful reminders of the strength of our God.

"They did not thirst when He led them through the deserts; He made water flow for them from the Rock; He split the rock and water gushed out." Isaiah 48:21

"You also, like living stones, are being built into a spiritual house... For as the scripture says, 'See, I lay a stone in Zion, a chosen and precious cornerstone, and the one who trusts in him will never be put to shame.' Now to you who believe, this stone is precious. But to those who do not believe, 'The stone the builder rejected has become the cornerstone,' and 'A stone that causes men to stumble and a rock that makes them fall.'" 1 Peter 2:5-8a

In Genesis 28, Jacob slept with his head on a rock, and never looked at the world in the same way. There is a stone that causes us to stumble and does more good than harm. His name is Jesus. If His heart had hardened because of our mistakes would there be any chance for us? It is precisely that He loves us enough to be hurt by us that we have a chance to be saved. The same should be true for our hearts. The world will hurt us. With Christ it will not defeat us.

The hearts of stone I now collect remind me of the promises of God and from what the Lord has saved me. I tell others why I gather them so that they will know.

". . . if you hear His voice, do not harden your hearts as you did in the rebellion, during the time of testing in the desert." Hebrews 3:7b-8

Hearts of stone are not easy to find. When I find one, I hold it close and think of my God. I thank Him for saving me from myself. Each one is a reminder of blessing that He gave me instead of the burden, barrier or injury that I would have chosen. They are not planted in my garden, but gathered in a bowl, sitting on ledges, and leaning against my home. Most I give to others and use to share my love for my LORD.

"Above all else, guard your heart, for it affects everything you do."
Proverbs 4:23

I don't want to turn to stone. Like the velveteen rabbit, I want "to be real." I want a heart that can be broken and that can feel hope and notice the most insignificant tear. Because tears mean something.

There are good things that can come from our past, from our pain. Our memories can help us notice when others are cementing themselves in place; What is happening when they are shutting themselves off. We are able to show them the garden instead of the grave. They help us embrace, cold veiled statues of people who now stand where we once stood.

"... My heart is poured out on the ground because my people are destroyed."
Lamentations 2:11

A heart of flesh bleeds, the ache then allow tears held back by cemented dams to flow. Those tears, like torrents have the power to wash away the hurt when God becomes our Rock.

"Pour out your heart like water in the presence of the Lord." Lamentations 2:19

A heart of flesh rekindles hope. Hope for another meeting with our Creator and hope to know and be known by One willing to look us in the eyes, who lets us believe again that some part of us is beautiful and can grow good things.

I hope I will always keep looking for my hearts of stone. I hope that you will join me in doing so. My collection waxes and wanes. I have forgotten from where most of them came. But I try to remember those I have given away and those to whom I have given them. I give them to my friends to remind them they are treasured. I have left them on the graves of those I miss or shared them with teenage girls going through struggles I have survived. I hope they are stepping stones, reminding them of the way to come back when they need them. Letting go of parts of my collections and hoping the one who picks them up again will find the same promise in the word of God. Because of them, I write this to help another find a reason for hope. I wrote it to point them to the cornerstone that is their hope.

"I will give thanks, for you answered me; you have become my salvation. The stone the builders rejected has become the cornerstone; The Lord has done this and it is marvelous in our eyes." Psalms 118:21-23

Ordinary things surround us every day. Some are useless. Some are treasure. It is stored and accumulated. The collections that pile up in our hearts may be used to build shelters, quiet patios away from the world, or to border blessings. Our strength and imagination set the only limits to the potential good that can be formed from them. Build what ignites joy from the gifts you harvest.

What makes something a treasure is the value we attribute to it. The God of the universe calls,

"The Lord your God has chosen you out of all the peoples on the face of the earth to be his people, His treasured possession." Deuteronomy 7:6

He gives us one another so that our hearts are kept open and able to be broken. Sometimes he plows. Other times He cuts away, and sometimes he plants. Hardship and heart-ache teach us the value of real treasure. In brokenness, we understand what matters most. Only because Christ's heart was broken and freely opened do we have access to the treasures of heaven.

C. S. Lewis said, "Of all powers, Love is the most powerful and most powerless. It is the most powerful, because it alone can conquer that final and most impregnable stronghold which is the human heart. It is most powerless, because it can do nothing except by consent."

The rock we live on is a hard and cold place with a rough exterior. We can believe promises, like the one in Ezekiel, are real because of Jesus Christ, His death and resurrection. He is the one whose heart was broken for us and He is the one who can restore the heart we were born with. The stone rolled away from the grave of Jesus let those looking for Him see that no stone can hold Him in or keep Him out. The same is true for the stone caves and prisons of the heart.

The stone that beat within my chest broke and was made new. I have tried to store and accumulate what God treasures since then. My heart has since been filled with treasure beyond measure and my joy is made complete cultivating, planting and harvesting life's precious gifts.

Have you seen a heart of stone restored to a heart of flesh?

Has a boulder ever stood in your way and eventually you thanked God for it?

Have you ever stared God in the face when He has asked you a question?

GARDEN
Always

Concerning The End...Before We Begin...

"He stilled the storm to a whisper; the waves of the sea were hushed. They were glad when it grew calm, and He guided them to their desired haven." Psalm 107:29-30

It has not happened often that I have come to the end of fall and thought my garden ready for December's chill. Most years snow blankets projects still incomplete in my yard. But at some point gardening must cease and no amount of trying stops winter coming.

When the ground freezes, spring can seem very far away. The leaves fall and tender perennials collapse to the ground. Anything unfinished and hidden by their foliage becomes painfully obvious. Other times too quickly a snowy blanket covers what I long to confront and hides it from my sight. Yet, the world is out of my control and beyond my strength to cultivate. Leaving things alone and keeping my hands off is a real struggle, especially, if I can see and even touch what is wrong.

"In the depth of winter, I finally learned that within me there lay an invincible summer." - Albert Camus

Winter is as inevitable in life as it is in the calendar that governs our years. The temperature drops around us; it hurts to touch anything and if we do, we only succeed in making things worse. There will be blizzards that obscure our ability to identify what is wrong. Winds that will never return what they have taken. And places that mimic Narnia governed by a white witch, where the world seems an eternal winter.

He sends his command to the earth; his word runs swiftly. He spreads the snow like wool and scatters the frost like ashes. He hurls down his hail like pebbles. Who can withstand his icy blast? He sends his word and melts them; he stirs up his breezes, and the waters flow."
Psalm 147:15-18

We have come a long way from Eden. It was a perfect place. It was a perfect moment that should have lasted, but a fallen world rushed in and the blessing was broken. Eden ended. No place on earth has been like it since. No matter how much we want places to stay at their peak, they don't. Flowers die away, weed seeds blow in and the weather is beyond our control. The world pulls and pushes us away. Everything changes. Everything dies. After all...we are human...

"Fire burns in front of them and flames follow after them. Ahead of them the land lies as beautiful as the Garden of Eden. Behind them is nothing but desolation, not one thing escapes." Joel 2:3

I wonder if Adam and Eve stood outside of Eden's boundaries hoping to wake from a bad dream or if they ran as fast and as long as they could away from it. Every man since has cried himself to sleep at some point and woke with the feeling he had lost something. Then recognized that he is lost!!!

"A man's own folly ruins his life, yet his heart rages against the Lord." Proverbs 19:3

From the first taste of sin Adam and Eve mistook their enemy. The Lord, whom formed their innocent hearts, became the God from whom they hide their guilt. He had to seek them. "But the LORD called to them, "Where are you?" Genesis 3:9 The truth is our Creator has been seeking us ever since. He has never given up. He never ceases to water us and garden around us.

"Sing about a fruitful vineyard: I, the LORD, watch over it; I water it continually. I guard it day and night so that no one may harm it."
Isaiah 27:3

Gardens are quiet places, but they have a lot to say about His unchanging, never-ending love. Blossoms and leaves shiver with ardent joy for their Creator. White flowers reflecting moonlight and swirling autumn leaves whisper His gentle patience. Reflection pools bring peace to us for this reason. The truths gardens echo and the sweet fragrances they release send ripples ringing around their boundaries to herald that undying passion. Every rushing waterfall and twinkle of light through the trees is a reminder of His overflowing love and outstretched arms.

"The grasslands of the desert overflow; the hills are clothed with gladness. The meadows are covered with flocks and the valleys are mantled in grain; they shout for joy and sing." Psalms 65: 12-13

However, most people have lost contact with the genuineness of a garden. Power tools have eased the curse of the ground, but have done little to cultivate God's kingdom. Our technology, like the tower of Babel, has taken the place of relationships. We find answers on the internet and fail to remember The Omniscient God.

"This only have I found: God made man upright, but men have gone in search of many schemes." Ecclesiastes 7:29

The world will swallow up, frustrate, and change the gifts of God when it can. Gardening will remind us,

. . . This is what the Sovereign LORD says: Although I sent them far away among the nations and scattered them among the countries, yet for a little while I have been a sanctuary for them. . .'"
Ezekiel 11:16

Gardening is simple and purposeful tasks that never end. Those little tasks change the world even if only a small corner of it at a time. At times it is little more than providing for basic needs. It adapts to harsh temperature, water tables and ever changing seasons. It is not always enjoyable, nor without cost, but gardening is always meaningful.

. . . The vineyard of the LORD ALMIGHTY is (my life) and (you) are the garden of His delight." Isaiah 5:7

"The world is full of obvious things which nobody by chance ever observes." S.H. Keep that in mind as you sort through this new way of approaching life. We want to be do great things with our time on earth, and with the heart of gardeners we can make the world a better place. My hope is that this journal will help others take up their role as gardeners and restore the relationships of the world's gardeners with the Master of gardens.

"The one who sows to please the spirit, from the spirit will reap eternal life." Galatians 6:8

When we enter a garden, it is a reminder of all He desires to restore before and within us. But, not all of us get to walk in garden paths every day. Not all of us have access to quiet places, oasis in the deserts, vacations in paradise, but we all have a God who knows what oppresses us and steals our peace. "When I tried to understand all this, it was oppressive to me, till I entered the sanctuary of God: then I understood …" Psalms 73:16-17a. We will never remake Eden. However, our efforts can focus on creating a sanctuary where we meet Him and invite His care over us. We can fill our lives with things from the Creator of all good things, and water friendships and dig deep into relationship with our God. We can cultivate lives, like gardens, that celebrate His providence and draw others to it.

When we traveled out west I learned that mountains make their own weather. It can feel like we are always climbing, always waiting for the sun to rise. We push ourselves to watch the sun touch the next ridge. Get up as early as possible so we can capture on a camera a fraction of the wonder there to be explored. Some clouds are good, but not too many. Then at the last minute it rains.

Harsh season don't last forever. They just keep us out for a while. In the darkest, longest, and coldest of winters, spring violets wait beneath the snow. In the hottest, driest, and harshest of places seeds wait for water. We are not without power in the winter or desert places to make a difference in the world around us. Even when distance and time separate us from personal contact there is hope. When there is no garden, and when winter and drought creep in, prayer is a greenhouse. A season we invite Him to work when we are not able. We may even get to plant seeds and keep warm as we learn to rest and hope for tomorrow.

A child of God in harsh seasons must shift his focus. He wields prayer when all hope is gone. Often the hardest thing to do is pray. We want to get our hands dirty. Feeling helpless is worse than being windburnt, hypothermic, and exhausted even on our day off. We want quick solutions and exact directions

to the point of contentment. The last thing we can bare is to stand by as what we love is blanketed in trouble or wait as a solid crust thickens between us. Worse still is watching what we love crumble under an oppressive heat as we long for water ourselves. We want anything but to be powerless to stop it. It is at this point we cry for help, desperate for His answer. We meet our prayer lives or become stiff, still and cold ourselves. Our hopes dry up and wind blows them away like dust.

God changes everything. He draws us close and wields us in the harshest of seasons.

"The Lord says, "Then I will heal you... and my love will know no bounds, for my anger will be gone forever. I will be to Israel like refreshing dew from heaven. It will blossom like the lily; it will send roots deep into the soil like a cedar in Lebanon. Its branches will spread out like those of the beautiful olive trees, as fragrant as the cedar forests of Lebanon... I am the one who looks after you and cares for you. I am like a tree that is always green, giving my fruit to you all through the year."
Hosea 14:4-7

In simple truths, we can find tranquility. The Bible, of course, is our clearest understanding of the Lord, but the thing the Lord used to draw me to himself was a garden.

"He reached down from on high and took hold of me; He drew me out of deep waters" 2 Samuel 22:17

I am confident He is using it to draw others to Himself as well.

J.I.Packer says:"So we are cruel to ourselves if we try to live in this world without knowing about whose world it is, and who runs it. The world becomes a strange, mad, painful place and life in it a disappointing business for those who do not know about God. Disregard the study of God and you sentence yourself to stumble and blunder through life blindfolded, as it were, with no sense of direction and no understanding of what surrounds you. This way you can waste your life and lose your soul." My hope is that others will recognize a love story that began in Eden through this journal. And that they will begin to collect their own bits and pieces of evidence. Our lives can be a heaven like Eden for those looking for an oasis. They can be places to cultivate a child's heart or sanctuaries where others find shelter and see glimpses of their Creator.

"What you are is God's gift to you. What you become is your gift to God."- Hans Urs von Balthasar

We never know what can grow from unidentified seeds planted in any given one of us. We will not get to see all that is grown from what our lives plant. So be patient, keep planting flowers, learning the names of all that delights you and taking note of extra care needs.

"... This is what the kingdom of God is like. A man scatters seed on the ground. Night and day, whether he sleeps or gets up, the seed sprouts and grows, though he does not know how. All by itself the soil produces grain..." Mark 4:26-28

Dare we imagine what treasure can be cultivated? Through the gift of gardening, we wake the parts of the world that sleep and are deaf, blind to a God that proclaims His unending love daily from the heavens above us and from the ground beneath us. Our investments will only remain gardens so long as we continue to take time to maintain them. What God calls us to garden is worth the price of commitment!!!

"They will be called oaks of righteousness a planting of the Lord for the display of his splendor." Isaiah 61:3b

"Forever, for always, no matter what. . ."

GARDEN
Always

For more information, please checkout the below sites:

Instagram: melanie_joyb

The Rusted Root
Webite: www.therustedroot.com
Instagram: 518therustedroot
Facebook: the rusted root, 518 Hoffman Road
Phone: 814-233-5890

07/24/2006

"Meaning is lost
without wonder."
- Ravi Zacharias